GEOMETRY
REGENTS REVIEW

Published by
TOPICAL REVIEW BOOK COMPANY
P. O. Box 328
Onsted, MI 49265-0328

GEOMETRY REGENTS REVIEW

Published by
TOPICAL REVIEW BOOK COMPANY
P. O. Box 328
Onsted, MI 49265-0328
www.topicalrbc.com

EXAM	PAGE
Reference Sheet	i
2008 – Fall Sampler	1
Test 1	10
June 2009	20
August 2009	29
January 2010	38
June 2010	48
August 2010	59

Geometry Reference Sheet

Volume	Cylinder	$V = Bh$ where B is the area of the base
	Pyramid	$V = \frac{1}{3}Bh$ where B is the area of the base
	Right Circular Cone	$V = \frac{1}{3}Bh$ where B is the area of the base
	Sphere	$V = \frac{4}{3}\pi r^3$

Lateral Area (L)	Right Circular Cylinder	$L = 2\pi rh$
	Right Circular Cone	$L = \pi rl$ where l is the slant height

Surface Area	Sphere	$SA = 4\pi r^2$

Note: A graphing calculator, a straightedge (ruler), and compass must be available for you to use while taking this examination.

GEOMETRY
2008 – Fall Sampler
Part I

Answer all 28 questions in this part. Each correct answer will receive 2 credits. No partial credit will be allowed. For each question, write in the space provided the numeral preceding the word or expression that best completes the statement or answers the question.

1. Isosceles trapezoid $ABCD$ has diagonals $\overline{AC}$ and $\overline{BD}$. If $AC = 5x + 13$ and $BD = 11x - 5$, what is the value of x?
(1) 28   (2) $10\frac{3}{4}$ (3) 3   (4) $\frac{1}{2}$

1 _____

2. What is the negation of the statement "The Sun is shining"?
(1) It is cloudy.   (3) It is not raining.
(2) It is daytime.   (4) The Sun is not shining.

2 _____

3. Triangle ABC has vertices $A(1, 3)$, $B(0, 1)$, and $C(4, 0)$. Under a translation, A', the image point of A, is located at $(4, 4)$. Under this same translation, point C' is located at
(1) $(7, 1)$ (2) $(5, 3)$ (3) $(3, 2)$ (4) $(1, -1)$

3 _____

4. The accompanying diagram shows the construction of the perpendicular bisector of $\overline{AB}$. Which statement is *not* true?
(1) $AC = CB$
(2) $CB = \frac{1}{2}AB$
(3) $AC = 2AB$
(4) $AC + CB = AB$

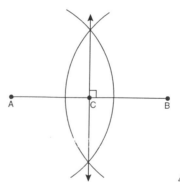

4 _____

5. Which graph could be used to find the solution to the following system of equations?
$y = -x + 2$
$y = x^2$

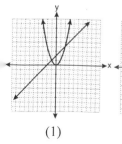
(1)

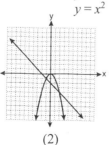
(2)

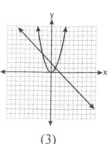
(3)

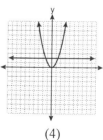
(4)

5 _____

GEOMETRY
2008 – Fall Sampler

6. Line *k* is drawn so that it is perpendicular to two distinct planes, *P* and *R*. What must be true about planes *P* and *R*?
(1) Planes *P* and *R* are skew.
(2) Planes *P* and *R* are parallel.
(3) Planes *P* and *R* are perpendicular.
(4) Plane *P* intersects plane *R* but is not perpendicular to plane *R*.   6 _____

7. The accompanying diagram illustrates the construction of $\overrightarrow{PS}$ parallel to $\overrightarrow{RQ}$ through point *P*. Which statement justifies this construction?
(1) m∠1 = m∠2
(2) m∠1 = m∠3
(3) $\overline{PR} \cong \overline{RQ}$
(4) $\overline{PS} \cong \overline{RQ}$

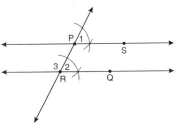

7 _____

8. The figure in the accompanying diagram is a triangular prism. Which statement must be true?
(1) $\overline{DE} \cong \overline{AB}$ (3) $\overline{AD} \parallel \overline{CE}$
(2) $\overline{AD} \cong \overline{BC}$ (4) $\overline{DE} \parallel \overline{BC}$

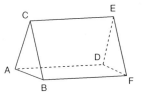

8 _____

9. The vertices of △*ABC* are *A*(−1, −2), *B*(−1, 2), and *C*(6, 0). Which conclusion can be made about the angles of △*ABC*?
(1) m∠*A* = m∠*B*   (3) m∠*ACB* = 90
(2) m∠*A* = m∠*C*   (4) m∠*ABC* = 60

9 _____

10. Given △*ABC* with base $\overline{AFEDC}$, median $\overline{BF}$, altitude $\overline{BD}$, and $\overline{BE}$ bisects ∠*ABC*, which conclusion is valid?
(1) ∠*FAB* ≅ ∠*ABF*
(2) ∠*ABF* ≅ ∠*CBD*
(3) $\overline{CE} \cong \overline{EA}$
(4) $\overline{CF} \cong \overline{FA}$

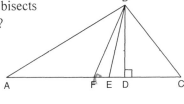

10 _____

11. In the accompanying diagram, circle *O* has a radius of 5, and *CE* = 2. Diameter $\overline{AC}$ is perpendicular to chord $\overline{BD}$ at *E*. What is the length of $\overline{BD}$?
(1) 12   (3) 8
(2) 10   (4) 4

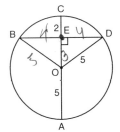

11 _____

GEOMETRY
2008 – Fall Sampler

12. What is the equation of a line that passes through the point (−3, −11) and is parallel to the line whose equation is $2x - y = 4$?

(1) $y = 2x + 5$
(2) $y = 2x - 5$
(3) $y = \frac{1}{2}x + \frac{25}{2}$
(4) $y = -\frac{1}{2}x - \frac{25}{2}$

12 _____

13. Line segment AB has endpoints $A(2, -3)$ and $B(-4, 6)$. What are the coordinates of the midpoint of $\overline{AB}$?

(1) $(-2, 3)$
(2) $(-1, 1\frac{1}{2})$
(3) $(-1, 3)$
(4) $(3, 4\frac{1}{2})$

13 _____

14. What are the center and radius of a circle whose equation is $(x - A)^2 + (y - B)^2 = C$?
(1) center = (A, B); radius = C
(2) center = $(-A, -B)$; radius = C
(3) center = (A, B); radius = $\sqrt{C}$
(4) center = $(-A, -B)$; radius = $\sqrt{C}$

14 _____

15. A rectangular prism has a volume of $3x^2 + 18x + 24$. Its base has a length of $x + 2$ and a width of 3. Which expression represents the height of the prism?
(1) $x + 4$
(2) $x + 2$
(3) 3
(4) $x^2 + 6x + 8$

15 _____

16. Lines k_1 and k_2 intersect at point E. Line m is perpendicular to lines k_1 and k_2 at point E. Which statement is always true?
(1) Lines k_1 and k_2 are perpendicular.
(2) Line m is parallel to the plane determined by lines k_1 and k_2.
(3) Line m is perpendicular to the plane determined by lines k_1 and k_2.
(4) Line m is coplanar with lines k_1 and k_2.

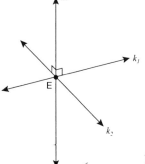

16 _____

17. In the accompanying diagram, $\overline{PS}$ is a tangent to circle O at point S, $\overline{PQR}$ is a secant, $PS = x$, $PQ = 3$, and $PR = x + 18$. What is the length of $\overline{PS}$?
(1) 6
(2) 9
(3) 3
(4) 27

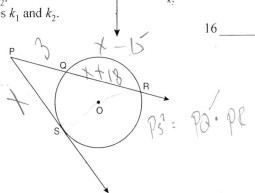

(Not drawn to scale)

17 _____

GEOMETRY
2008 – Fall Sampler

18. A polygon is transformed according to the rule: $(x, y) \to (x + 2, y)$. Every point of the polygon moves two units in which direction?
(1) up (2) down (3) left (4) right   18 _____

19. In the accompanying diagram of $\triangle ABC$, D is a point on $\overline{AB}$, $AC = 7$, $AD = 6$, and $BC = 18$. The length of $\overline{DB}$ could be
(1) 5   (3) 19
(2) 12   (4) 25

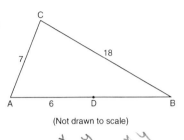
(Not drawn to scale)   19 _____

20. The diameter of a circle has endpoints at $(-2, 3)$ and $(6, 3)$. What is an equation of the circle?
(1) $(x - 2)^2 + (y - 3)^2 = 16$
(2) $(x - 2)^2 + (y - 3)^2 = 4$
(3) $(x + 2)^2 + (y + 3)^2 = 16$
(4) $(x + 2)^2 + (y + 3)^2 = 4$ 20 _____

21. In the accompanying diagram of $\triangle PRT$, Q is a point on $\overline{PR}$, S is a point on $\overline{TR}$, $\overline{QS}$ is drawn, and $\angle RPT \cong \angle RSQ$. Which reason justifies the conclusion that $\triangle PRT \sim \triangle SRQ$?
(1) AA   (3) SAS
(2) ASA (4) SSS

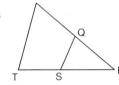
     21 _____

22. The lines $3y + 1 = 6x + 4$ and $2y + 1 = x - 9$ are
(1) parallel   (3) the same line
(2) perpendicular   (4) neither parallel nor perpendicular   22 _____

23. The endpoints of $\overline{AB}$ are $A(3, 2)$ and $B(7, 1)$. If $\overline{A''B''}$ is the result of the transformation of $\overline{AB}$ under $D_2 \circ T_{-4, 3}$ what are the coordinates of A'' and B'' ?
(1) A'' $(-2, 10)$ and B'' $(6, 8)$
(2) A'' $(-1, 5)$ and B'' $(3, 4)$
(3) A'' $(2, 7)$ and B'' $(10, 5)$
(4) A'' $(14, -2)$ and B'' $(22, -4)$ 23 _____

24. In the accompanying diagram, circle A and circle B are shown. What is the total number of lines of tangency that are common to circle A and circle B?
(1) 1   (3) 3
(2) 2   (4) 4

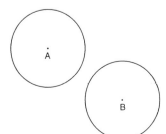
     24 _____

GEOMETRY
2008 – Fall Sampler

25. In which triangle do the three altitudes intersect outside the triangle?
(1) a right triangle (3) an obtuse triangle
(2) an acute triangle (4) an equilateral triangle   25 _____

26. Two triangles are similar, and the ratio of each pair of corresponding sides is 2 : 1. Which statement regarding the two triangles is *not* true?
(1) Their areas have a ratio of 4 : 1.
(2) Their altitudes have a ratio of 2 : 1.
(3) Their perimeters have a ratio of 2 : 1.
(4) Their corresponding angles have a ratio of 2 : 1.   26 _____

27. What is the measure of an interior angle of a regular octagon?
(1) 45°   (2) 60°   (3) 120°   (4) 135°   27 _____

28. What is the slope of a line perpendicular to the line whose equation is $5x + 3y = 8$?
(1) $\frac{5}{3}$ (2) $\frac{3}{5}$ (3) $-\frac{3}{5}$ (4) $-\frac{5}{3}$ 28 _____

Part II
Answer all 6 questions in this part. Each correct answer will receive 2 credits. Clearly indicate the necessary steps, including appropriate formula substitutions, diagrams, graphs, charts, etc. For all questions in this part, a correct numerical answer with no work shown will receive only 1 credit. [12]

29. In the diagram below of right triangle ACB, altitude $\overline{CD}$ intersects $\overline{AB}$ at D. If $AD = 3$ and $DB = 4$, find the length of $\overline{CD}$ in simplest radical form.

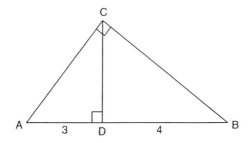

GEOMETRY
2008 – Fall Sampler

30. The vertices of △ABC are A(3, 2), B(6, 1), and C(4, 6). Identify and graph a transformation of △ABC such that its image, △A′B′C′, results in $\overline{AB} \parallel \overline{A'B'}$.

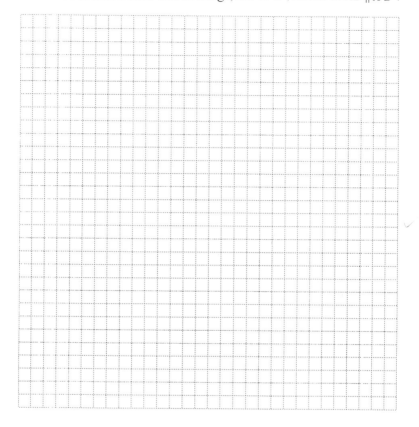

31. The endpoints of $\overline{PQ}$ are P(−3, 1) and Q(4, 25). Find the length of $\overline{PQ}$.

GEOMETRY
2008 – Fall Sampler

32. Using a compass and straightedge, construct the bisector of the angle shown below. [*Leave all construction marks.*]

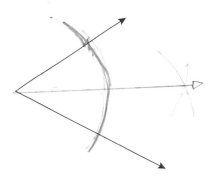

33. The volume of a cylinder is 12,566.4 cm³. The height of the cylinder is 8 cm. Find the radius of the cylinder to the *nearest tenth of a centimeter*.

34. Write a statement that is logically equivalent to the statement "If two sides of a triangle are congruent, the angles opposite those sides are congruent."

Identify the new statement as the converse, inverse, or contrapositive of the original statement.

GEOMETRY
2008 – Fall Sampler
Part III

Answer all 3 questions in this part. Each correct answer will receive 4 credits. Clearly indicate the necessary steps, including appropriate formula substitutions, diagrams, graphs, charts, etc. For all questions in this part, a correct numerical answer with no work shown will receive only 1 credit. [12]

35. On the set of axes below, graph and label $\triangle DEF$ with vertices at $D(-4, -4)$, $E(-2, 2)$, and $F(8, -2)$.

If G is the midpoint of $\overline{EF}$ and H is the midpoint of $\overline{DF}$, state the coordinates of G and H and label each point on your graph.

Explain why $\overline{GH} \parallel \overline{DE}$

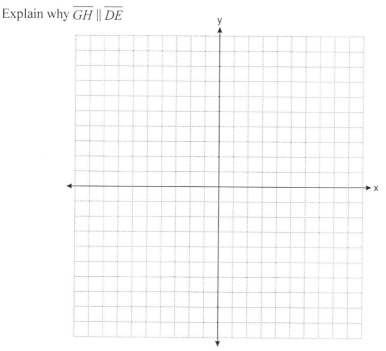

36. In the diagram below of circle O, chords $\overline{DF}$, $\overline{DE}$, $\overline{FG}$, and $\overline{EG}$ are drawn such that $m\widehat{DF} : m\widehat{FE} : m\widehat{EG} : m\widehat{GD} = 5 : 2 : 1 : 7$. Identify one pair of inscribed angles that are congruent to each other and give their measure.

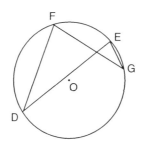

GEOMETRY
2008 – Fall Sampler

37. A city is planning to build a new park. The park must be equidistant from school A at $(3, 3)$ and school B at $(3, -5)$. The park also must be exactly 5 miles from the center of town, which is located at the origin on the coordinate graph. Each unit on the graph represents 1 mile.

On the set of axes below, sketch the compound loci and label with an **X** all possible locations for the new park.

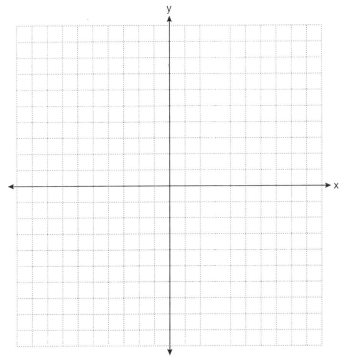

Part IV

Answer the question in this part. The correct answer will receive 6 credits. Clearly indicate the necessary steps, including appropriate formula substitutions, diagrams, graphs, charts, etc. For the question in this part, a correct numerical answer with no work shown will receive only 1 credit. [6]

38. In the accompanying diagram, quadrilateral $ABCD$ is inscribed in circle O, $\overline{AB} \parallel \overline{DC}$, and diagonals $\overline{AC}$ and $\overline{BD}$ are drawn.
Prove that $\triangle ACD \cong \triangle BDC$.

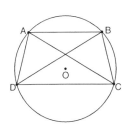

GEOMETRY
TEST 1
Part I

Answer all 28 questions in this part. Each correct answer will receive 2 credits. No partial credit will be allowed. For each question, write in the space provided the numeral preceding the word or expression that best completes the statement or answers the question.

1. Which is the equation of a line that is parallel to the line $y = \frac{1}{2}x + 3$?
(1) $y = 2x + 3$
(2) $y = \frac{1}{2}x + 4$
(3) $y = -2x + 1$
(4) $y = -\frac{1}{2}x + 3$

1 _____

2. Which of the following sets of numbers could *not* be the measures of the three sides of a triangle?
(1) 1, 7, 7 (2) 2, 3, 5 (3) 7, 9, 15 (4) 4, 4, 7

2 _____

3. Which is the equation of a line that passes through the point (1, 0) and is perpendicular to the line $4y = -2x + 8$?
(1) $y = x + 8$ (2) $y = -x - 4$ (3) $y = -2x - 1$ (4) $y = 2x - 2$

3 _____

4. The volume of a right circular cone is 150π cubic centimeters. Find its altitude in centimeters if the radius of the base is 5 cm.
(1) 2 (2) 6 (3) 18 (4) 45

4 _____

5. In a regular pyramid, which of the following statements is always true?
(1) The base is a rhombus.
(2) The base is a square.
(3) The edges of the pyramid are all congruent.
(4) The lateral faces of the pyramid are congruent isosceles triangles.

5 _____

6. What is the equation of circle A in the accompanying diagram?
(1) $(x - 1)^2 + (y - 3)^2 = 16$
(2) $(x + 1)^2 + (y + 3)^2 = 16$
(3) $(x + 3)^2 - (y + 1)^2 = 16$
(4) $(x + 1)^2 - (y + 3)^2 = 16$

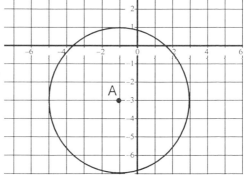

6 _____

7. Under a reflection over the line $y = 0$, which is *not* preserved?
(1) orientation (2) size (3) shape (4) perimeter

7 _____

GEOMETRY
TEST 1

8. What is the center of a circle with the following equation?
$$(x-1)^2 + (y+2)^2 = 49$$
(1) (1, 2) (2) (1, –2) (3) (–1, –2) (4) (–1, 2)

8 _____

9. Which transformation describes the accompanying figures?
(1) $R_{x\ axis}$
(2) $R_{y\ axis}$
(3) $R_{y\ =\ x}$
(4) $R_{y\ =\ -x}$

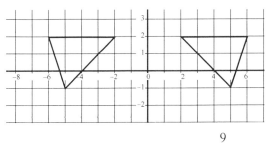

9 _____

10. Rhombus ABCD is shown to the right. Which can *not* be the length of diagonal $\overline{AC}$?
(1) 24 m   (3) 14 m
(2) 18 m   (4) 11 m

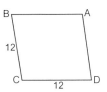

10 _____

11. Parallel lines m and n are located 10 meters apart. How many points are equidistant from m and n and also 5 meters from Point C?

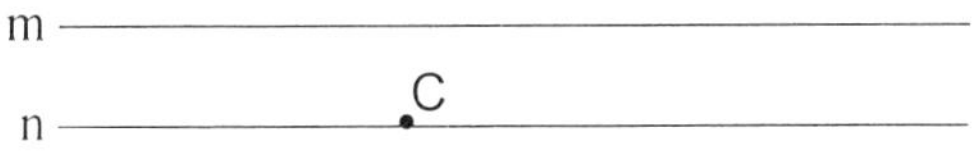

(1) 1   (2) 2   (3) 3   (4) 4

11 _____

12. Line n and line m are coplanar and intersect at point A. Line r is perpendicular to both lines n and m at A. Describe the relationship between lines n, m, and r.
(1) Line r is parallel to the plane formed by lines n and m.
(2) Line r is perpendicular to the plane formed by lines n and m.
(3) A right triangle is formed by lines m, n, and r.
(4) All the intersections of lines n, m, and r form right angles.

12 _____

13. $\overline{AC}$ is tangent to circle O at B. $\overline{BD}$ is a chord. Arc BD = 110. What is the measure of $\angle ABD$?
(1) 110   (3) 200
(2) 125   (4) 250

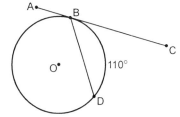

13 _____

14. Point H is located at (–5, 7). What are the coordinates of the image of Point H after a 180° rotation about the origin?
(1) (–7, –5)   (2) (5, –7)   (3) (7, 5)   (4) (–5, 7)

14 _____

GEOMETRY
TEST 1

15. Line *m* is perpendicular to plane *A*. How many planes can contain line *m* and also be perpendicular to plane *A*?
(1) 1　　(2) 2　　(3) 0　　(4) An infinite number

15 _____

16. The sum of the exterior angles added to the sum of the interior angles of a polygon must be
(1) less than 360°
(2) equal to 360°
(3) greater than or equal to 540°
(4) greater than or equal to 720°

16 _____

17. What is the length of a line segment with the endpoints (1, 5) and (3, 2)?
(1) $\sqrt{13}$　　(2) $\sqrt{5}$　　(3) $\sqrt{30}$　　(4) $\sqrt{65}$

17 _____

18. What is the negation of the statement, "The bus was traveling at 55 miles per hour."
(1) The bus was traveling at 50 miles per hour.
(2) The bus was traveling at 65 miles per hour.
(3) The bus was not traveling at 55 miles per hour.
(4) The bus was stopped.

18 _____

19. What is the relationship of the two lines with the equations $-3y = x - 6$ and $y = 3x + 1$?
(1) parallel
(2) perpendicular
(3) intersecting
(4) none of the above

19 _____

20. In right triangle *ABC*, altitude $\overline{BD}$ is drawn. $AD = 3$ and $DC = 27$. What is the length of $\overline{BD}$?
(1) 9　　(3) 15
(2) 12　　(4) 24

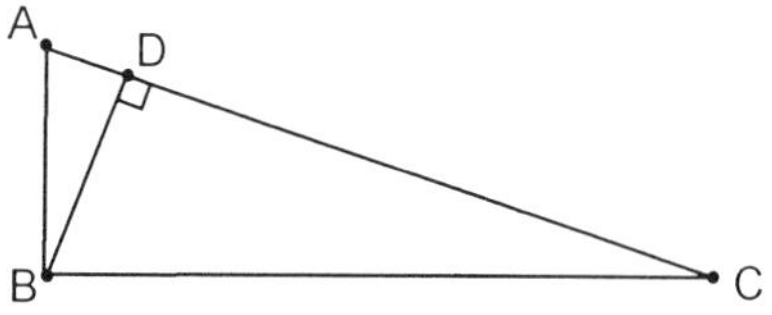

20 _____

21. In the accompanying diagram of circle *O*, chords $\overline{AB}$ and $\overline{CD}$ intersect at *E*. If $AE = 2$, $EB = 10$, $DE = x$ and $CE = x - 1$ what is the value of *x*?
(1) 5.5　　(3) 2.5
(2) 5　　(4) 0.25

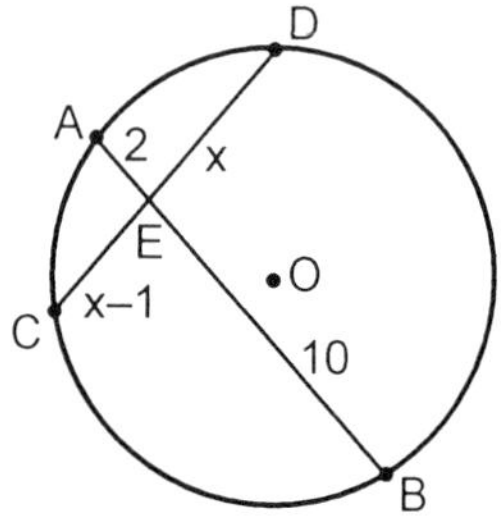

21 _____

22. The legs of a right triangle measure $2\sqrt{3}$ and 5. What is the length of the hypotenuse?
(1) $\sqrt{31}$　　(2) $\sqrt{37}$　　(3) $\sqrt{43}$　　(4) $\sqrt{17}$

22 _____

GEOMETRY
TEST 1

23. In △ABC, medians $\overline{AD}$, $\overline{BF}$, and $\overline{CE}$ are drawn and intersect at G. EG = 2. What is the length of $\overline{GC}$?
(1) 1
(2) 2
(3) 4
(4) 6

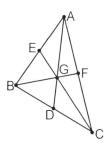

23 _____

24. Parallelogram ABCD has diagonals $\overline{AC}$ and $\overline{BD}$. Which of the following statements is always true?
(1) $\frac{1}{2}(AC)(BD)$ = Area of ABCD
(2) AC = BD
(3) △ABC ≅ △CDA
(4) △ABD ≅ △ABC

24 _____

25. Two triangles are similar. The lengths of the sides of the smaller triangle are 4, 6, 7. The longest side of the larger triangle is 21. What is the perimeter of the larger triangle?
(1) 17 (2) 18 (3) 42 (4) 51

25 _____

26. In circle O, ABC is a secant, $\overline{AE}$ is tangent to the circle at E. What is the length of AE?
(1) $5\sqrt{2}$ (3) 5
(2) $2\sqrt{5}$ (4) 7.5

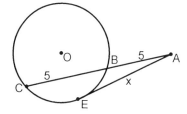

26 _____

27. Which is the equation of a circle with a center of (5, 2) and a radius of 10?
(1) $(x + 5)^2 - (y + 2)^2 = 10$
(2) $(x - 5)^2 + (y - 2)^2 = 10$
(3) $(x + 5)^2 - (y + 2)^2 = 100$
(4) $(x - 5)^2 + (y - 2)^2 = 100$

27 _____

28. In triangle ABC, $\overline{ED}$ connects the midpoints of $\overline{AB}$ and $\overline{AC}$. The length of $\overline{ED}$ is 8. What is the length of $\overline{BC}$?
(1) 4
(2) 12
(3) 16
(4) 20

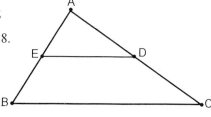

28 _____

14

GEOMETRY
TEST 1
Part II

Answer all 6 questions in this part. Each correct answer will receive 2 credits. Clearly indicate the necessary steps, including appropriate formula substitutions, diagrams, graphs, charts, etc. For all questions in this part, a correct numerical answer with no work shown will receive only 1 credit. [12]

29. Find the coordinates of the midpoint of the segment joining $(-2, 4)$ and $(8, 6)$.

30. Quadrilateral $RSTU$ has vertices of $R(-3, -2)$, $S(-6, -2)$, $T(-7, -4)$ and $U(-2, -4)$. Draw and state the coordinates of the image, $R'S'T'U'$, under a glide reflection where $T_{0,4}$ and r_{y-axis} are performed.

GEOMETRY
TEST 1

31. Construct a bisector for the angle shown below. (Leave all construction marks)

32. Given this *true* statement: If Joe lives in New York State, then he starts school in September. Write the contrapositive of this statement *and* state whether it is true or false.

33. What is the radius of a sphere that has a surface area of 144π?

GEOMETRY
TEST 1

34. Write the equation of the locus of points two units from (–3, 4). The use of the coordinate grid is optional.

GEOMETRY
TEST 1
Part III

Answer all 3 questions in this part. Each correct answer will receive 4 credits. Clearly indicate the necessary steps, including appropriate formula substitutions, diagrams, graphs, charts, etc. For all questions in this part, a correct numerical answer with no work shown will receive only 1 credit. [12]

35. Solve this system of equations graphically:

$$y = -x^2 + 2x + 4$$
$$y = x - 2$$

36. $\overline{AB}$ and $\overline{CD}$ are intersected by $\overline{EF}$ at G and H as shown in the diagram. $\angle EGA = 3x - 6$ and $\angle CHF = 2x + 4$. If $x = 6$, determine if $\overline{AB} \parallel \overline{CD}$ and justify your answer.

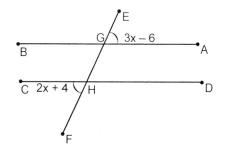

37. The perimeter of a square is 72. Express the length of a diagonal of the square in simplest radical form.

GEOMETRY
TEST 1
Part IV

Answer the question in this part. The correct answer will receive 6 credits. Clearly indicate the necessary steps, including appropriate formula substitutions, diagrams, graphs, charts, etc. For the question in this part, a correct numerical answer with no work shown will receive only 1 credit. [6]

38. Given: Quadrilateral $QRST$
    $\overline{QR} \parallel \overline{TS}$, $\overline{TS} \perp \overline{SQ}$, $\overline{TQ} \cong \overline{RS}$

    Prove: $\overline{QR} \cong \overline{TS}$

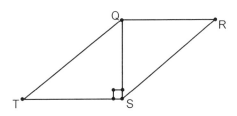

GEOMETRY
June 2009

Part I

Answer all 28 questions in this part. Each correct answer will receive 2 credits. No partial credit will be allowed. For each question, record your answer in the space provided.

1. Juliann plans on drawing △*ABC*, where the measure of ∠*A* can range from 50° to 60° and the measure of ∠*B* can range from 90° to 100°. Given these conditions, what is the correct range of measures possible for ∠*C*?
   (1) 20° to 40°   (2) 30° to 50°   (3) 80° to 90°   (4) 120° to 130°   1 _____

2. In the accompanying diagram of △*ABC* and △*DEF*, $\overline{AB} \cong \overline{DE}$, ∠*A* ≅ ∠*D*, and ∠*B* ≅ ∠*E*. Which method can be used to prove △*ABC* ≅ △*DEF*?
   (1) SSS   (3) ASA
   (2) SAS   (4) HL   2 _____

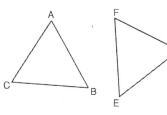

3. In the accompanying diagram, under which transformation will △*A'B'C'* be the image of △*ABC*?
   (1) rotation   (3) translation
   (2) dilation   (4) glide reflection   3 _____

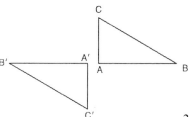

4. The lateral faces of a regular pyramid are composed of
   (1) squares   (3) congruent right triangles
   (2) rectangles   (4) congruent isosceles triangles   4 _____

5. Point *A* is located at (4, −7). The point is reflected in the *x*-axis. Its image is located at
   (1) (−4, 7)   (2) (−4, −7)   (3) (4, 7)   (4) (7, −4)   5 _____

6. In the accompanying diagram of circle *O*, chords $\overline{AB}$ and $\overline{CD}$ are parallel, and $\overline{BD}$ is a diameter of the circle. If m*AD* = 60, what is m∠*CDB*?
   (1) 20   (3) 60
   (2) 30   (4) 120   6 _____

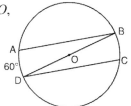

GEOMETRY
June 2009

7. What is an equation of the line that passes through the point (−2, 5) and is perpendicular to the line whose equation is $y = \frac{1}{2}x + 5$?
(1) $y = 2x + 1$ (2) $y = -2x + 1$ (3) $y = 2x + 9$ (4) $y = -2x - 9$ 7 _____

8. After a composition of transformations, the coordinates A(4, 2), B(4, 6), and C(2, 6) become A″(−2, −1), B″(−2, −3), and C″(−1, −3), as shown on the accompanying set of axes.

Which composition of transformations was used?
(1) $R_{180°} \circ D_2$ (3) $D_{\frac{1}{2}} \circ R_{180°}$
(2) $R_{90°} \circ D_2$ (4) $D_{\frac{1}{2}} \circ R_{90°}$ 8 _____

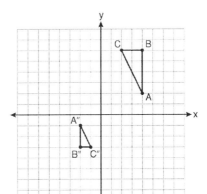

9. In an equilateral triangle, what is the difference between the sum of the exterior angles and the sum of the interior angles?
(1) 180° (2) 120° (3) 90° (4) 60°   9 _____

10. What is an equation of a circle with its center at (−3, 5) and a radius of 4?
(1) $(x - 3)^2 + (y + 5)^2 = 16$ (3) $(x - 3)^2 + (y + 5)^2 = 4$
(2) $(x + 3)^2 + (y - 5)^2 = 16$ (4) $(x + 3)^2 + (y - 5)^2 = 4$ 10 _____

11. In △ABC, m∠A = 95, m∠B = 50, and m∠C = 35. Which expression correctly relates the lengths of the sides of this triangle?
1) AB < BC < CA   (3) AC < BC < AB
2) AB < AC < BC   (4) BC < AC < AB   11 _____

12. In a coordinate plane, how many points are both 5 units from the origin and 2 units from the x-axis?
1) 1 (2) 2 (3) 3 (4) 4   12 _____

13. What is the contrapositive of the statement, "If I am tall, then I will bump my head"?
1) If I bump my head, then I am tall.
2) If I do not bump my head, then I am tall.
3) If I am tall, then I will not bump my head.
4) If I do not bump my head, then I am not tall.   13 _____

14. In the accompanying diagram of △ABC, Jose found centroid P by constructing the three medians. He measured $\overline{CF}$ and found it to be 6 inches. If $PF = x$, which equation can be used to find x? 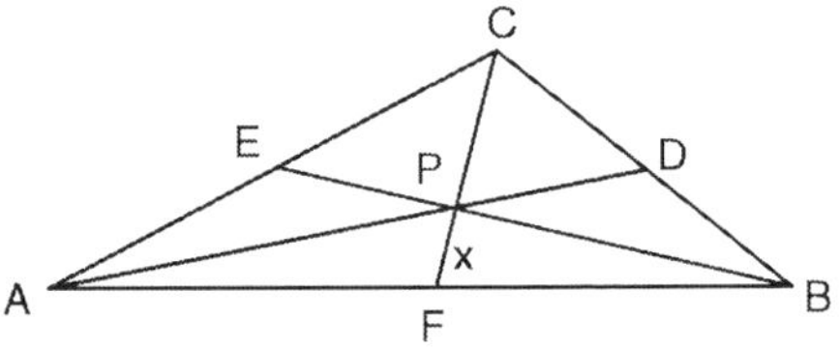
(1) $x + x = 6$
(2) $2x + x = 6$
(3) $3x + 2x = 6$
(4) $x + \frac{2}{3}x = 6$

14 _____

15. In the accompanying diagram, the length of the legs $\overline{AC}$ and $\overline{BC}$ of right triangle ABC are 6 cm and 8 cm, respectively. Altitude $\overline{CD}$ is drawn to the hypotenuse of △ABC. What is the length of $\overline{AD}$ to the *nearest tenth of a centimeter*? 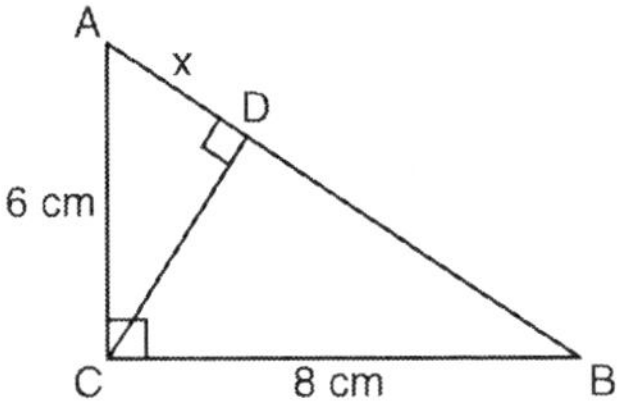
(1) 3.6   (2) 6.0   (3) 6.4   (4) 4.0

15 _____

16. In the accompanying diagram, tangent $\overline{AB}$ and secant $\overline{ACD}$ are drawn to circle O from an external point A, $AB = 8$, and $AC = 4$. What is the length of $\overline{CD}$? 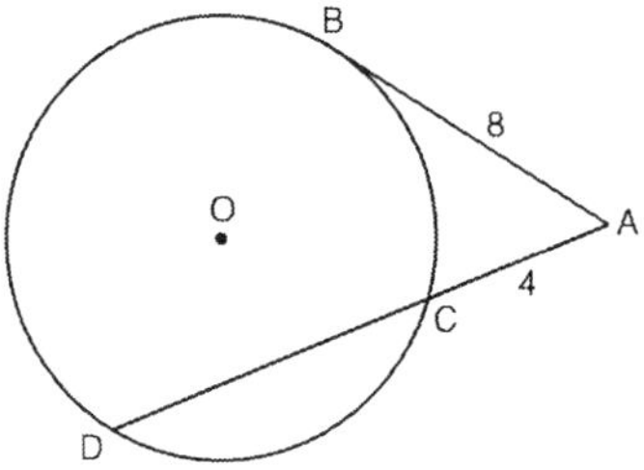
(1) 16   (2) 13   (3) 12   (4) 10

16 _____

17. In the accompanying diagram of △ABC and △EDC, $\overline{AE}$ and $\overline{BD}$ intersect at C, and $\angle CAB \cong \angle CED$. Which method can be used to show that △ABC must be similar to △EDC? 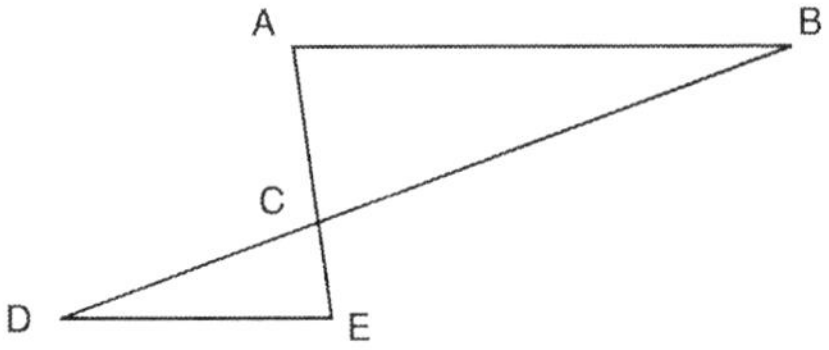
(1) SAS   (2) AA   (3) SSS   (4) HL

17 _____

18. Point P is on line m. What is the total number of planes that are perpendicular to line m and pass through point P?
(1) 1   (2) 2   (3) 0   (4) infinite

18 _____

GEOMETRY
June 2009

19. Square *LMNO* is shown in the accompanying diagram. What are the coordinates of the midpoint of diagonal $\overline{LN}$?

(1) $(4\frac{1}{2}, -2\frac{1}{2})$ (3) $(-2\frac{1}{2}, 3\frac{1}{2})$

(2) $(-3\frac{1}{2}, 3\frac{1}{2})$ (4) $(-2\frac{1}{2}, 4\frac{1}{2})$

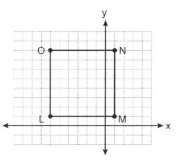

19 _____

20. Which graph represents a circle with the equation $(x - 5)^2 + (y + 1)^2 = 9$?

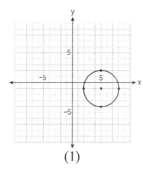
(1)

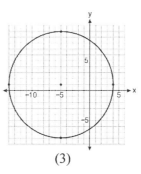
(3)

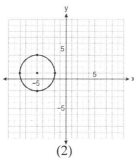
(2)

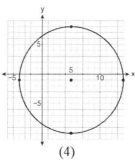
(4)

20 _____

21. In the accompanying diagram, a right circular cone has a diameter of 8 inches and a height of 12 inches. What is the volume of the cone to the *nearest cubic inch*?

(1) 201   (3) 603
(2) 481   (4) 804

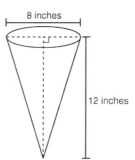

21 _____

GEOMETRY
June 2009

22. A circle is represented by the equation $x^2 + (y + 3)^2 = 13$. What are the coordinates of the center of the circle and the length of the radius?
(1) (0, 3) and 13
(2) (0, 3) and $\sqrt{13}$
(3) (0, −3) and 13
(4) (0, −3) and $\sqrt{13}$

22 _____

23. Given the system of equations: $y = x^2 - 4x$
$x = 4$
The number of points of intersection is
(1) 1   (2) 2   (3) 3   (4) 0

23 _____

24. Side $\overline{PQ}$ of $\triangle PQR$ is extended through Q to point T. Which statement is *not* always true?
(1) $m\angle RQT > m\angle R$
(2) $m\angle RQT > m\angle P$
(3) $m\angle RQT = m\angle P + m\angle R$
(4) $m\angle RQT > m\angle PQR$

24 _____

25. Which illustration shows the correct construction of an angle bisector?

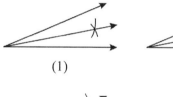

(1)   (3)

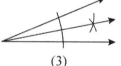

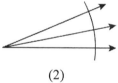

(2)   (4)

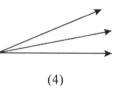

25 _____

26. Which equation represents a line perpendicular to the line whose equation is $2x + 3y = 12$?
(1) $6y = -4x + 12$
(2) $2y = 3x + 6$
(3) $2y = -3x + 6$
(4) $3y = -2x + 12$

26 _____

27. In $\triangle ABC$, point D is on $\overline{AB}$, and point E is on $\overline{BC}$ such that $\overline{DE} \parallel \overline{AC}$. If $DB = 2$, $DA = 7$, and $DE = 3$, what is the length of $\overline{AC}$?
(1) 8   (2) 9   (3) 10.5   (4) 13.5

27 _____

28. In three-dimensional space, two planes are parallel and a third plane intersects both of the parallel planes. The intersection of the planes is a
(1) plane
(2) point
(3) pair of parallel lines
(4) pair of intersecting lines

28 _____

GEOMETRY
June 2009
Part II

Answer all 6 questions in this part. Each correct answer will receive 2 credits. Clearly indicate the necessary steps, including appropriate formula substitutions, diagrams, graphs, charts, etc. For all questions in this part, a correct numerical answer with no work shown will receive only 1 credit. All answers should be written in pen, except for graphs and drawings, which should be done in pencil.

29. In the accompanying diagram of $\triangle ABC$, $AB = 10$, $BC = 14$, and $AC = 16$. Find the perimeter of the triangle formed by connecting the midpoints of the sides of $\triangle ABC$.

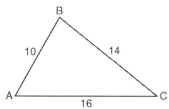

30. Using a compass and straightedge, construct a line that passes through point P and is perpendicular to line m. [Leave all construction marks.]

31. Find an equation of the line passing through the point (5, 4) and parallel to the line whose equation is $2x + y = 3$.

32. The length of $\overline{AB}$ is 3 inches. On the diagram below, sketch the points that are equidistant from A and B and sketch the points that are 2 inches from A. Label with an **X** all points that satisfy *both* conditions.

33. Given: Two is an even integer or three is an even integer. Determine the truth value of this disjunction. Justify your answer.

34. In the diagram below, $\triangle ABC \sim \triangle EFG$, $m\angle C = 4x + 30$, and $m\angle G = 5x + 10$. Determine the value of x.

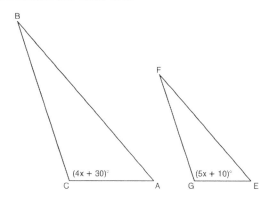

GEOMETRY
June 2009
Part III

Answer all 3 questions in this part. Each correct answer will receive 4 credits. Clearly indicate the necessary steps, including appropriate formula substitutions, diagrams, graphs, charts, etc. For all questions in this part, a correct numerical answer with no work shown will receive only 1 credit. All answers should be written in pen, except for graphs and drawings, which should be done in pencil.

35. In the diagram below, circles X and Y have two tangents drawn to them from external point T. The points of tangency are C, A, S, and E. The ratio of TA to AC is 1:3. If $TS = 24$, find the length of $\overline{SE}$.

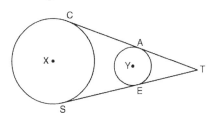

(Not drawn to scale)

36. Triangle ABC has coordinates $A(-6, 2)$, $B(-3, 6)$, and $C(5, 0)$. Find the perimeter of the triangle. Express your answer in simplest radical form. [The use of the grid below is optional.]

GEOMETRY
June 2009

37. The coordinates of the vertices of parallelogram $ABCD$ are $A(-2, 2)$, $B(3, 5)$, $C(4, 2)$, and $D(-1, -1)$. State the coordinates of the vertices of parallelogram $A''B''C''D''$ that result from the transformation $r_{y\text{-axis}} \circ T_{2,-3}$. [The use of the set of axes below is optional.]

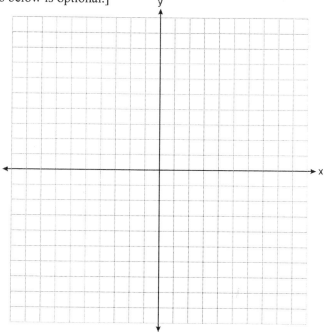

Part IV

Answer the question in this part. A correct answer will receive 6 credits. Clearly indicate the necessary steps, including appropriate formula substitutions, diagrams, graphs, charts, etc. A correct numerical answer with no work shown will receive only 1 credit. The answer should be written in pen.

38. Given: $\triangle ABC$ and $\triangle EDC$, C is the midpoint of $\overline{BD}$ and $\overline{AE}$
Prove: $\overline{AB} \parallel \overline{DE}$

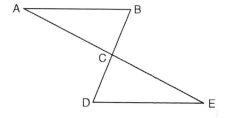

GEOMETRY
August 2009
Part I

Answer all 28 questions in this part. Each correct answer will receive 2 credits. No partial credit will be allowed. For each question, write in the space provided the numeral preceding the word or expression that best completes the statement or answers the question. [56]

1. Based on the accompanying diagram, which statement is true?

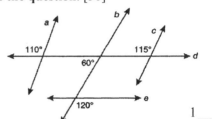

(1) $a \parallel b$ (3) $b \parallel c$
(2) $a \parallel c$ (4) $d \parallel e$

1 _____

2. The diagram below shows the construction of the bisector of $\angle ABC$. Which statement is *not* true?

(1) $m\angle EBF = \frac{1}{2} m\angle ABC$

(2) $m\angle DBF = \frac{1}{2} m\angle ABC$

(3) $m\angle EBF = m\angle ABC$

(4) $m\angle DBF = m\angle EBF$

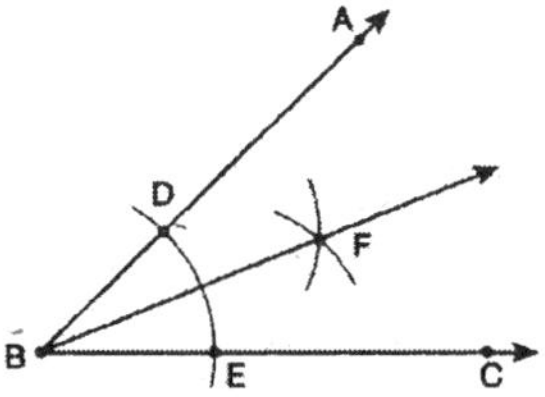

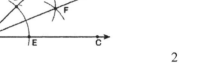

2 _____

3. In the accompanying diagram of $\triangle ABC$, $\overline{AB} \cong \overline{AC}$. The measure of $\angle B$ is $40°$. What is the measure of $\angle A$?
(1) $40°$ (3) $70°$
(2) $50°$ (4) $100°$

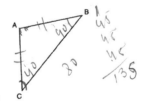

3 _____

4. In the accompanying diagram of circle O, chord $\overline{CD}$ is parallel to diameter $\overline{AOB}$ and $m\widehat{AC} = 30$. What is $m\widehat{CD}$?
(1) 150 (3) 100
(2) 120 (4) 60

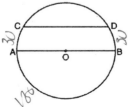

4 _____

GEOMETRY
August 2009

5. In the accompanying diagram of trapezoid $ABCD$, diagonals $\overline{AC}$ and $\overline{BD}$ intersect at E and $\triangle ABC \cong \triangle DCB$. Which statement is true based on the given information?

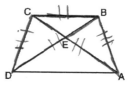

(1) $\overline{AC} \cong \overline{BC}$
(2) $\overline{CD} \cong \overline{AD}$
(3) $\angle CDE \cong \angle BAD$
(4) $\angle CDB \cong \angle BAC$ ✓

5 _____

6. Which transformation produces a figure similar but *not* congruent to the original figure?
(1) $T_{1,3}$ →
(2) $D_{\frac{1}{2}}$
(3) $R_{90°}$
(4) $r_{y=x}$ ✓

6 _____

7. In the accompanying diagram of parallelogram $ABCD$ with diagonals $\overline{AC}$ and $\overline{BD}$, m$\angle 1 = 45$ and m$\angle DCB = 120$. What is the measure of $\angle 2$?

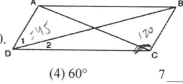

✓ (1) 15°   (2) 30°   (3) 45°   (4) 60°

7 _____

8. On the accompanying set of axes, Geoff drew rectangle $ABCD$. He will transform the rectangle by using the translation $(x, y) \to (x + 2, y + 1)$ and then will reflect the translated rectangle over the x-axis. What will be the area of the rectangle after these transformations?

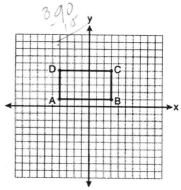

✓ (1) exactly 28 square units
(2) less than 28 square units
(3) greater than 28 square units
(4) It cannot be determined from the information given.

8 _____

9. What is the equation of a line that is parallel to the line whose equation is $y = x + 2$?
(1) $x + y = 5$
(2) $2x + y = -2$
(3) $y - x = -1$ ✓
(4) $y - 2x = 3$

9 _____

10. The endpoints of $\overline{CD}$ are $C(-2, -4)$ and $D(6, 2)$. What are the coordinates of the midpoint of $\overline{CD}$?
(1) (2,3)   ✓ (2) (2, -1)   (3) (4, -2)   (4) (4, 3)

10 _____

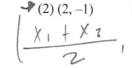

Center at origin
$x^2 + y^2 = r^2$

GEOMETRY
August 2009

not on origin
$(x-k)^2 + (y-h)^2 = r^2$ 31

11. What are the center and the radius of the circle whose equation is $(x - 3)^2 + (y + 3)^2 = 36$?
(1) center = (3, –3); radius = 6   (3) center = (3, –3); radius = 36
(2) center = (–3, 3); radius = 6   (4) center = (–3, 3); radius = 36   11 _____

12. Given the equations: $y = x^2 - 6x + 10$
    $y + x = 4$ — get y by itself.
What is the solution to the given system of equations?
(1) (2, 3)   (2) (3, 2)   (3) (2, 2) and (1, 3)   **(4)** (2, 2) and (3, 1)   12 _____

13. The diagonal $\overline{AC}$ is drawn in parallelogram ABCD. Which method can *not* be used to prove that $\triangle ABC \cong \triangle CDA$?
(1) SSS   (2) SAS   **(3)** SSA   (4) ASA   13 _____

14. In the accompanying diagram, line k is perpendicular to plane P at point T. Which statement is true?
(1) Any point in plane P also will be on line k.
(2) Only one line in plane P will intersect line k.
(3) All planes that intersect plane P will pass through T.
(4) Any plane containing line k is perpendicular to plane P.   14 _____

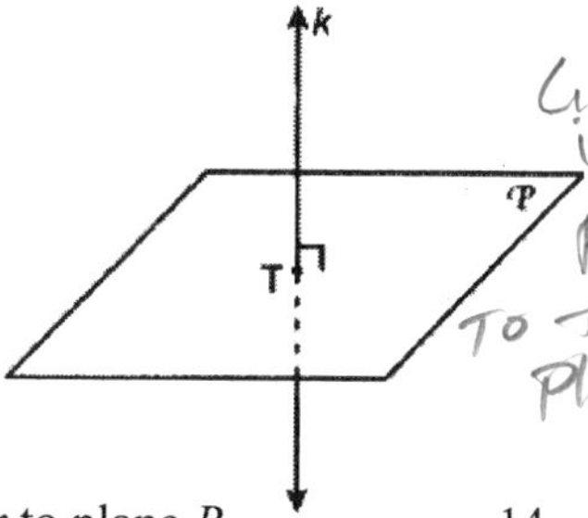

Line k is perpendicular to the plane.

15. In the accompanying diagram, which transformation was used to map $\triangle ABC$ to $\triangle A'B'C'$? 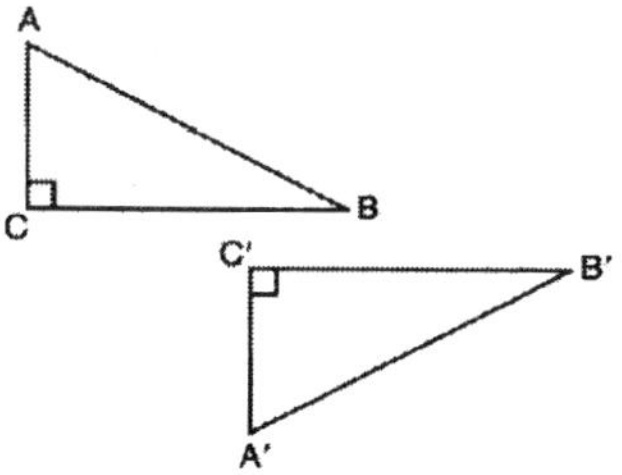
(1) dilation - bigger
(2) rotation - turn
(3) reflection - opposite
(4) glide reflection   15 _____

16. Which set of numbers represents the lengths of the sides of a triangle?
(1) {5, 18, 13}   **(2)** {6, 17, 22}   (3) {16, 24, 7}   (4) {26, 8, 15}   16 _____

17. What is the slope of a line perpendicular to the line whose equation is $y = -\frac{2}{3}x - 5$? $y = mx + b$ slope - negative reciprocal
(1) $-\frac{3}{2}$ (2) $-\frac{2}{3}$ (3) $\frac{2}{3}$ **(4)** $\frac{3}{2}$ switch the slope   17 _____

18. A quadrilateral whose diagonals bisect each other and are perpendicular is a
(1) rhombus   (2) rectangle   (3) trapezoid   (4) parallelogram   18 _____

GEOMETRY
August 2009

19. If the endpoints of $\overline{AB}$ are $A(-4, 5)$ and $B(2, -5)$, what is the length of $\overline{AB}$?
(1) $2\sqrt{34}$ (2) 2   (3) $\sqrt{61}$ (4) 8   19 _____

20. In the accompanying diagram of $\triangle ACT$, D is the midpoint of $\overline{AC}$, O is the midpoint of $\overline{AT}$, and G is the midpoint of $\overline{CT}$. If $AC = 10$, $AT = 18$, and $CT = 22$, what is the perimeter of parallelogram $CDOG$?
(1) 21   (2) 25   (3) 32   (4) 40   20 _____

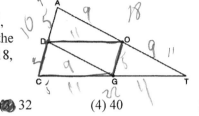

21. Which equation represents circle K shown in the accompanying graph?
(1) $(x + 5)^2 + (y - 1)^2 = 3$
(2) $(x + 5)^2 + (y - 1)^2 = 9$
(3) $(x - 5)^2 + (y + 1)^2 = 3$
(4) $(x - 5)^2 + (y + 1)^2 = 9$

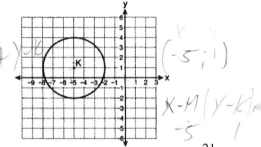

21 _____

22. In the accompanying diagram of right triangle ACB, altitude $\overline{CD}$ is drawn to hypotenuse $\overline{AB}$. If $AB = 36$ and $AC = 12$, what is the length of $\overline{AD}$?
(1) 32   (3) 3
(2) 6   (4) 4

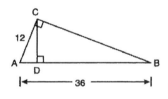

22 _____

23. In the accompanying diagram of circle O, chord $\overline{AB}$ intersects chord $\overline{CD}$ at E, $DE = 2x + 8$, $EC = 3$, $AE = 4x - 3$, and $EB = 4$. What is the value of x? 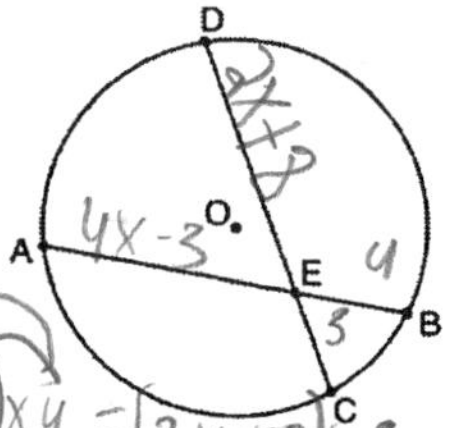
(1) 1   (3) 5
(2) 3.6   (4) 10.25

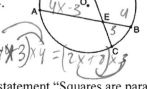

23 _____

24. What is the negation of the statement "Squares are parallelograms"?
(1) Parallelograms are squares.
(2) Parallelograms are not squares.
(3) It is not the case that squares are parallelograms.
(4) It is not the case that parallelograms are squares.

24 _____

GEOMETRY
August 2009

25. The accompanying diagram shows the construction of the center of the circle circumscribed about $\triangle ABC$. This construction represents how to find the intersection of
(1) the angle bisectors of $\triangle ABC$
(2) the medians to the sides of $\triangle ABC$
(3) the altitudes to the sides of $\triangle ABC$
(4) the perpendicular bisectors of the sides of $\triangle ABC$

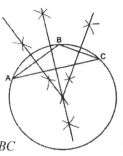

25 _____

26. A right circular cylinder has a volume of 1,000 cubic inches and a height of 8 inches. What is the radius of the cylinder to the *nearest tenth of an inch*?
(1) 6.3   (2) 11.2   (3) 19.8   (4) 39.8   26 _____

27. If two different lines are perpendicular to the same plane, they are
(1) collinear   (2) coplanar   (3) congruent   (4) consecutive   27 _____

28. How many common tangent lines can be drawn to the two externally tangent circles shown to the right? 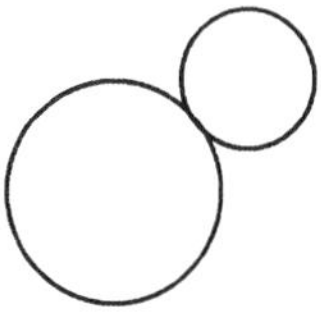
(1) 1   (3) 3
(2) 2   (4) 4

28 _____

Part II

Answer all 6 questions in this part. Each correct answer will receive 2 credits. Clearly indicate the necessary steps, including appropriate formula substitutions, diagrams, graphs, charts, etc. For all questions in this part, a correct numerical answer with no work shown will receive only 1 credit. [12]

29. In the accompanying diagram of isosceles trapezoid $DEFG$, $\overline{DE} \parallel \overline{GF}$, $DE = 4x - 2$, $EF = 3x + 2$, $FG = 5x - 3$, and $GD = 2x + 5$. Find the value of x.

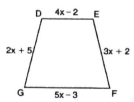

GEOMETRY
August 2009

30. A regular pyramid with a square base is shown in the accompanying diagram. A side, s, of the base of the pyramid is 12 meters, and the height, h, is 42 meters. What is the volume of the pyramid in cubic meters?

31. Write an equation of the line that passes through the point (6, −5) and is parallel to the line whose equation is $2x - 3y = 11$.

32. Using a compass and straightedge, construct the angle bisector of $\angle ABC$ shown below. [Leave all construction marks.]

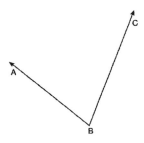

33. The degree measures of the angles of $\triangle ABC$ are represented by x, $3x$, and $5x - 54$. Find the value of x.

GEOMETRY
August 2009

34. In the diagram below of $\triangle ABC$ with side $\overline{AC}$ extended through D, m$\angle A$ = 37 and m$\angle BCD$ = 117. Which side of $\triangle ABC$ is the longest side? Justify your answer.

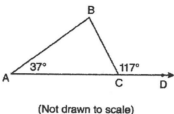

(Not drawn to scale)

Part III

Answer all 3 questions in this part. Each correct answer will receive 4 credits. Clearly indicate the necessary steps, including appropriate formula substitutions, diagrams, graphs, charts, etc. For all questions in this part, a correct numerical answer with no work shown will receive only 1 credit. [12]

35. Write an equation of the perpendicular bisector of the line segment whose endpoints are (–1, 1) and (7, –5). [The use of the grid below is optional]

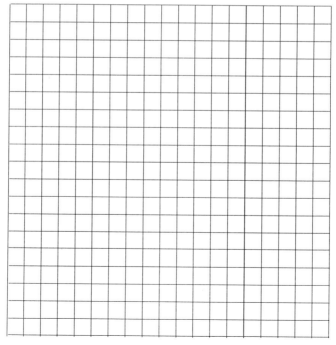

36. On the set of axes below, sketch the points that are 5 units from the origin and sketch the points that are 2 units from the line $y = 3$. Label with an **X** all points that satisfy *both* conditions.

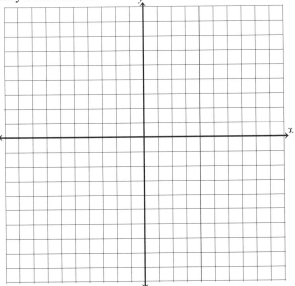

37. Triangle *DEG* has the coordinates $D(1, 1)$, $E(5, 1)$, and $G(5, 4)$. Triangle *DEG* is rotated 90° about the origin to form $\triangle D'E'G'$. On the grid below, graph and label $\triangle DEG$ and $\triangle D'E'G'$. State the coordinates of the vertices D', E', and G'. Justify that this transformation preserves distance.

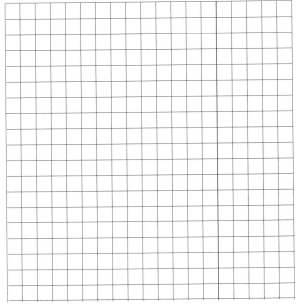

GEOMETRY
August 2009
Part IV

Answer the question in this part. The correct answer will receive 6 credits. Clearly indicate the necessary steps, including appropriate formula substitutions, diagrams, graphs, charts, etc. For the question in this part, a correct numerical answer with no work shown will receive only 1 credit. [6]

38. Given: Quadrilateral $ABCD$, diagonal $\overline{AFEC}$, $\overline{AE} \cong \overline{FC}$, $\overline{BF} \perp \overline{AC}$, $\overline{DE} \perp \overline{AC}$, $\angle 1 \cong \angle 2$

    Prove: $ABCD$ is a parallelogram.

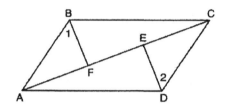

GEOMETRY
January 2010
Part I

Answer all 28 questions in this part. Each correct answer will receive 2 credits. No partial credit will be allowed. For each question, write in the space provided the numeral preceding the word or expression that best completes the statement or answers the question. [56]

1. In the accompanying diagram of trapezoid RSUT, $\overline{RS} \parallel \overline{TU}$, X is the midpoint of $\overline{RT}$, and V is the midpoint of $\overline{SU}$. If RS = 30 and XV = 44, what is the length of $\overline{TU}$?
   (1) 37   (2) 58   (3) 74   (4) 118   1 _____

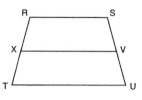

2. In △ABC, m∠A = x, m∠B = 2x + 2, and m∠C = 3x + 4. What is the value of x?
   (1) 29   (2) 31   (3) 59   (4) 61   2 _____

3. Which expression best describes the transformation shown in the accompanying diagram?
   (1) same orientation; reflection
   (2) opposite orientation; reflection
   (3) same orientation; translation
   (4) opposite orientation; translation

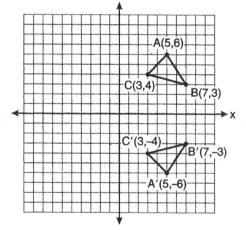

3 _____

4. Based on the accompanying construction, which statement must be true?
   (1) m∠ABD = $\frac{1}{2}$ m∠CBD
   (2) m∠ABD = m∠CBD
   (3) m∠ABD = m∠ABC
   (4) m∠CBD = $\frac{1}{2}$ m∠ABD

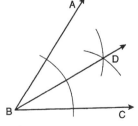

4 _____

GEOMETRY
January 2010

1. In the accompanying diagram, △ABC is inscribed in circle P. The distances from the center of circle P to each side of the triangle are shown. Which statement about the sides of the triangle is true?
(1) $AB > AC > BC$
(2) $AB < AC$ and $AC > BC$
(3) $AC > AB > BC$
(4) $AC = AB$ and $AB > BC$

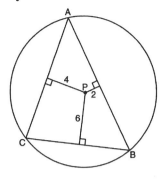

2. Which transformation is *not* always an isometry?
(1) rotation   (2) dilation   (3) reflection   (4) translation

3. In △ABC, $\overline{AB} \cong \overline{BC}$. An altitude is drawn from B to $\overline{AC}$ and intersects $\overline{AC}$ at D. Which statement is *not* always true?
(1) $\angle ABD \cong \angle CBD$
(2) $\angle BDA \cong \angle BDC$
(3) $\overline{AD} \cong \overline{BD}$
(4) $\overline{AD} \cong \overline{DC}$

4. In the accompanying diagram, tangent $\overline{PA}$ and secant $\overline{PBC}$ are drawn to circle O from external point P. If $PB = 4$ and $BC = 5$, what is the length of $\overline{PA}$?
(1) 20   (3) 8
(2) 9   (4) 6

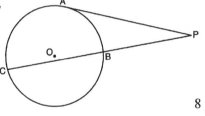

5. Which geometric principle is used to justify the accompanying construction? 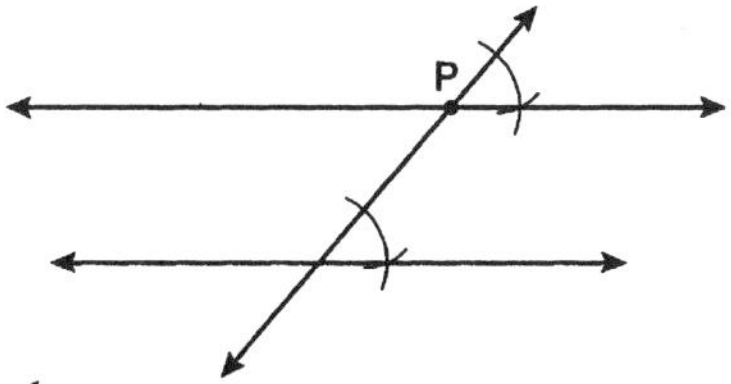
(1) A line perpendicular to one of two parallel lines is perpendicular to the other.
(2) Two lines are perpendicular if they intersect to form congruent adjacent angles.
(3) When two lines are intersected by a transversal and alternate interior angles are congruent, the lines are parallel.
(4) When two lines are intersected by a transversal and the corresponding angles are congruent, the lines are parallel.

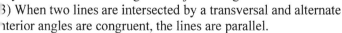

6. Which equation represents the circle whose center is (–2, 3) and whose radius is 5?
(1) $(x-2)^2 + (y+3)^2 = 5$
(2) $(x+2)^2 + (y-3)^2 = 5$
(3) $(x+2)^2 + (y-3)^2 = 25$
(4) $(x-2)^2 + (y+3)^2 = 25$

11. Towns *A* and *B* are 16 miles apart. How many points are 10 miles from town *A* and 12 miles from town *B*?
(1) 1 (2) 2 (3) 3 (4) 0   11 _____

12. Lines *j* and *k* intersect at point *P*. Line *m* is drawn so that it is perpendicular to lines *j* and *k* at point *P*. Which statement is correct?
(1) Lines *j* and *k* are in perpendicular planes.
(2) Line *m* is in the same plane as lines *j* and *k*.
(3) Line *m* is parallel to the plane containing lines *j* and *k*.
(4) Line *m* is perpendicular to the plane containing lines *j* and *k*.   12 _____

13. In the accompanying diagram of parallelogram *STUV*, $SV = x + 3$, $VU = 2x - 1$, and $TU = 4x - 3$. What is the length of $\overline{SV}$?
(1) 5   (3) 7
(2) 2   (4) 4

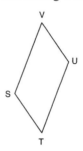

13 _____

14. Which equation represents a line parallel to the line whose equation is $2y - 5x = 10$?
(1) $5y - 2x = 25$ (3) $4y - 10x = 12$
(2) $5y + 2x = 10$ (4) $2y + 10x = 8$ 14 _____

15. In the accompanying diagram of circle *O*, chords $\overline{AD}$ and $\overline{BC}$ intersect at *E*, $m\widehat{AC} = 87$, and $m\widehat{BD} = 35$. What is the degree measure of $\angle CEA$?
(1) 87   (3) 43.5
(2) 61   (4) 26

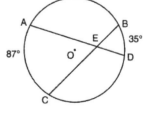

15 _____

16. In the accompanying diagram of $\triangle ADB$, $m\angle BDA = 90$, $AD = 5\sqrt{2}$, and $AB = 2\sqrt{15}$. What is the length of $\overline{BD}$?
(1) $\sqrt{10}$ (3) $\sqrt{50}$
(2) $\sqrt{20}$ (4) $\sqrt{110}$

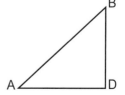

16 _____

17. What is the distance between the points $(-3, 2)$ and $(1, 0)$?
(1) $2\sqrt{2}$ (2) $2\sqrt{3}$ (3) $5\sqrt{2}$ (4) $2\sqrt{5}$ 17 _____

GEOMETRY
January 2010

18. What is an equation of the line that contains the point (3, –1) and is perpendicular to the line whose equation is $y = -3x + 2$?

(1) $y = -3x + 8$ (2) $y = -3x$ (3) $y = \frac{1}{3}x$ (4) $y = \frac{1}{3}x - 2$

18 _____

19. In the accompanying diagram, $\overline{SQ}$ and $\overline{PR}$ intersect at T, $\overline{PQ}$ is drawn, and $\overline{PS} \parallel \overline{QR}$. Which technique can be used to prove $\triangle PST \sim \triangle RQT$?
(1) SAS   (3) ASA
(2) SSS   (4) AA

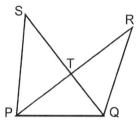

19 _____

20. The equation of a circle is $(x - 2)^2 + (y + 4)^2 = 4$. Which diagram is the graph of the circle?

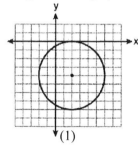
(1)

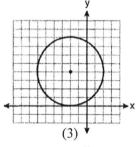
(3)

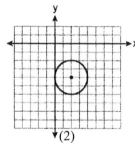
(2)

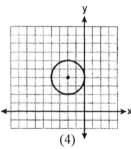
(4)

20 _____

21. In the accompanying diagram, $\triangle ABC$ is shown with $\overline{AC}$ extended through point D. If $m\angle BCD = 6x + 2$, $m\angle BAC = 3x + 15$, and $m\angle ABC = 2x - 1$, what is the value of x?
1) 12
2) $14\frac{10}{11}$
3) 16
4) $18\frac{1}{9}$

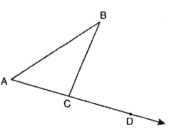

21 _____

22. Given △ABC ~ △DEF such that $\frac{AB}{DE} = \frac{3}{2}$. Which statement is *not* true?

(1) $\frac{BC}{EF} = \frac{3}{2}$ (3) $\frac{\text{area of } \triangle ABC}{\text{area of } \triangle DEF} = \frac{9}{4}$

(2) $\frac{m\angle A}{m\angle D} = \frac{3}{2}$ (4) $\frac{\text{perimeter of } \triangle ABC}{\text{perimeter of } \triangle DEF} = \frac{3}{2}$ 22 _____

23. The pentagon in the accompanying diagram is formed by five rays. What is the degree measure of angle *x*?
(1) 72   (3) 108
(2) 96   (4) 112

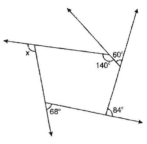

23 _____

24. Through a given point, *P*, on a plane, how many lines can be drawn that are perpendicular to that plane?
(1) 1   (2) 2   (3) more than 2   (4) none   24 _____

25. What is the slope of a line that is perpendicular to the line whose equation is $3x + 4y = 12$?
(1) $\frac{3}{4}$ (2) $-\frac{3}{4}$ (3) $\frac{4}{3}$ (5) $-\frac{4}{3}$ 25 _____

26. What is the image of point $A(4, 2)$ after the composition of transformations defined by $R_{90°} \circ r_{y=x}$?
(1) (–4, 2)   (2) (4, –2)   (3) (–4, –2)   (4) (2, –4)   26 _____

27. Which expression represents the volume, in cubic centimeters, of the cylinder represented in the accompanying diagram?
(1) 162π   (3) 972π
(2) 324π   (4) 3,888π

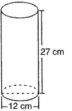

27 _____

28. What is the inverse of the statement "If two triangles are not similar, their corresponding angles are not congruent"?
(1) If two triangles are similar, their corresponding angles are not congruent.
(2) If corresponding angles of two triangles are not congruent, the triangles are not similar.
(3) If two triangles are similar, their corresponding angles are congruent.
(4) If corresponding angles of two triangles are congruent, the triangles are similar.   28 _____

GEOMETRY
January 2010
Part II

Answer all 6 questions in this part. Each correct answer will receive 2 credits. Clearly indicate the necessary steps, including appropriate formula substitutions, diagrams, graphs, charts, etc. For all questions in this part, a correct numerical answer with no work shown will receive only 1 credit. All answers should be written in pen, except for graphs and drawings, which should be done in pencil. [12]

29. In $\triangle RST$, m$\angle RST$ = 46 and $\overline{RS} \cong \overline{ST}$. Find m$\angle STR$.

30. Tim has a rectangular prism with a length of 10 centimeters, a width of 2 centimeters, and an unknown height. He needs to build another rectangular prism with a length of 5 centimeters and the same height as the original prism. The volume of the two prisms will be the same. Find the width, in centimeters, of the new prism.

31. In the diagram below of circle C, $\overline{QR}$ is a diameter, and $Q(1, 8)$ and $C(3.5, 2)$ are points on a coordinate plane. Find and state the coordinates of point R.

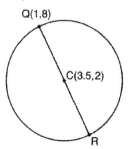

32. Using a compass and straightedge, and $\overline{AB}$ below, construct an equilateral triangle with all sides congruent to AB. [Leave all construction marks.]

33. In the diagram below of $\triangle ACD$, E is a point on $\overline{AD}$ and B is a point on $\overline{AC}$, such that $\overline{EB} \parallel \overline{DC}$. If $AE = 3$, $ED = 6$, and $DC = 15$, find the length of $\overline{EB}$.

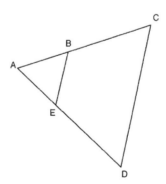

34. In the diagram below of $\triangle TEM$, medians $\overline{TB}$, $\overline{EC}$, and $\overline{MA}$ intersect at D, and $TB = 9$. Find the length of $\overline{TD}$.

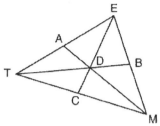

Part III

Answer all 3 questions in this part. Each correct answer will receive 4 credits. Clearly indicate the necessary steps, including appropriate formula substitutions, diagrams, graphs, charts, etc. For all questions in this part, a correct numerical answer with no work shown will receive only 1 credit. All answers should be written in pen, except for graphs and drawings, which should be done in pencil. [12]

35. In $\triangle KLM$, $m\angle K = 36$ and $KM = 5$. The transformation D_2 is performed on $\triangle KLM$ to form $\triangle K'L'M'$.

Find $m\angle K'$. Justify your answer.

Find the length of $\overline{K'M'}$. Justify your answer.

36. Given: JKLM is a parallelogram.
   $\overline{JM} \cong \overline{LN}$
   $\angle LMN \cong \angle LNM$
   Prove: JKLM is a rhombus.

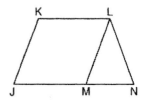

37. On the grid below, graph the points that are equidistant from both the x and y axes and the points that are 5 units from the origin. Label with an **X** all points that satisfy both conditions.

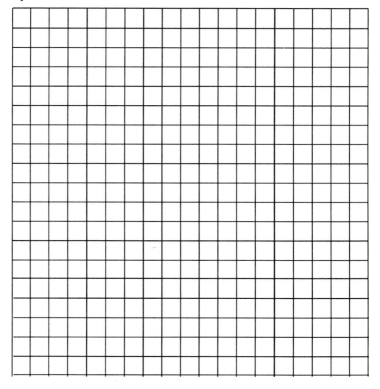

GEOMETRY
January 2010
Part IV

Answer the question in this part. A correct answer will receive 6 credits. Clearly indicate the necessary steps, including appropriate formula substitutions, diagrams, graphs, charts, etc. A correct numerical answer with no work shown will receive only 1 credit. The answer should be written in pen. [6]

38. On the set of axes below, solve the following system of equations graphically for all values of x and y.

$$y = (x - 2)^2 + 4$$
$$4x + 2y = 14$$

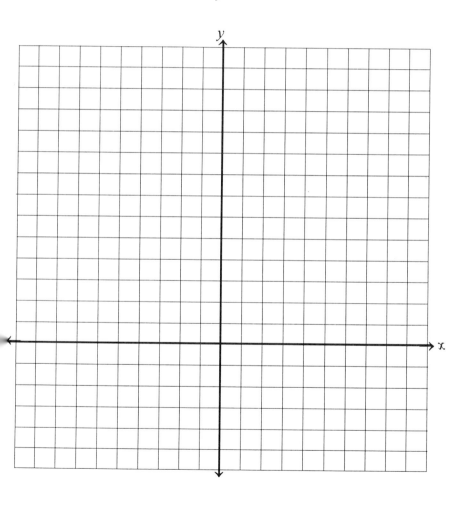

GEOMETRY
June 2010
Part I

Answer all 28 questions in this part. Each correct answer will receive 2 credits. No partial credit will be allowed. For each question, write on the space provided the numeral preceding the word or expression that best completes the statement or answers the question. [56]

1. In the accompanying diagram of circle O, chord $\overline{AB}$ ∥ chord $\overline{CD}$, and chord $\overline{CD}$ ∥ chord $\overline{EF}$. Which statement must be true?

(1) $\widehat{CE} \cong \widehat{DF}$

(2) $\widehat{AC} \cong \widehat{DF}$

(3) $\widehat{AC} \cong \widehat{CE}$

(4) $\widehat{EF} \cong \widehat{CD}$

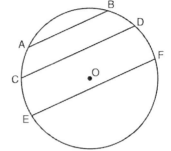

1 _____

2. What is the negation of the statement "I am not going to eat ice cream"?
(1) I like ice cream.
(2) I am going to eat ice cream.
(3) If I eat ice cream, then I like ice cream.
(4) If I don't like ice cream, then I don't eat ice cream.

2 _____

3. The accompanying diagram shows a right pentagonal prism.

Which statement is always true?

(1) $\overline{BC}$ ∥ $\overline{ED}$

(2) $\overline{FG}$ ∥ $\overline{CD}$

(3) $\overline{FJ}$ ∥ $\overline{IH}$

(4) $\overline{GB}$ ∥ $\overline{HC}$

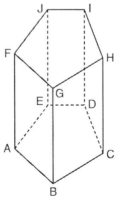

3 _____

4. In isosceles triangle ABC, $AB = BC$. Which statement will always be true?
(1) m∠B = m∠A (3) m∠A = m∠C
(2) m∠A > m∠B (4) m∠C < m∠B

4 _____

GEOMETRY
June 2010

5. The rectangle ABCD shown in the diagram below will be reflected across the x-axis.

What will *not* be preserved?
(1) slope of $\overline{AB}$
(2) parallelism of $\overline{AB}$ and $\overline{CD}$
(3) length of $\overline{AB}$
(4) measure of $\angle A$

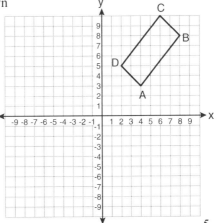

5 _____

6. A right circular cylinder has an altitude of 11 feet and a radius of 5 feet. What is the lateral area, in square feet, of the cylinder, to the *nearest tenth*?
(1) 172.7 (2) 172.8 (3) 345.4 (4) 345.6

6 _____

7. A transversal intersects two lines. Which condition would always make the two lines parallel?
(1) Vertical angles are congruent.
(2) Alternate interior angles are congruent.
(3) Corresponding angles are supplementary.
(4) Same-side interior angles are complementary.

7 _____

8. If the diagonals of a quadrilateral do *not* bisect each other, then the quadrilateral could be a
(1) rectangle (2) rhombus (3) square (4) trapezoid

8 _____

9. What is the converse of the statement "If Bob does his homework, then George gets candy"?
(1) If George gets candy, then Bob does his homework.
(2) Bob does his homework if and only if George gets candy.
(3) If George does not get candy, then Bob does not do his homework.
(4) If Bob does not do his homework, then George does not get candy.

9 _____

10. In $\triangle PQR$, $PQ = 8$, $QR = 12$, and $RP = 13$. Which statement about the angles of $\triangle PQR$ must be true?
(1) $m\angle Q > m\angle P > m\angle R$ (3) $m\angle R > m\angle P > m\angle Q$
(2) $m\angle Q > m\angle R > m\angle P$ (4) $m\angle P > m\angle R > m\angle Q$

10 _____

11. Given: $y = \frac{1}{4}x - 3$
$y = x^2 + 8x + 12$
In which quadrant will the graphs of the given equations intersect?
(1) I (2) II (3) III (4) IV   11 _____

12. Which diagram shows the construction of an equilateral triangle?

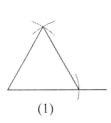   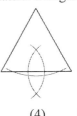

(1)   (2)   (3)   (4)   12 _____

13. Line segment AB is tangent to circle O at A. Which type of triangle is always formed when points A, B, and O are connected?
(1) right   (2) obtuse   (3) scalene   (4) isosceles   13 _____

14. What is an equation for the circle shown in the accompanying graph?
(1) $x^2 + y^2 = 2$
(2) $x^2 + y^2 = 4$
(3) $x^2 + y^2 = 8$
(4) $x^2 + y^2 = 16$

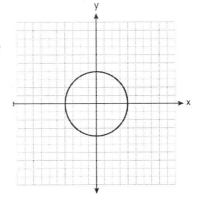

14 _____

15. Which transformation can map the letter **S** onto itself?
(1) glide reflection   (3) line reflection
(2) translation   (4) rotation   15 _____

16. In isosceles trapezoid $ABCD$, $\overline{AB} \cong \overline{CD}$. If $BC = 20$, $AD = 36$, and $AB = 17$, what is the length of the altitude of the trapezoid?
(1) 10   (2) 12   (3) 15   (4) 16   16 _____

17. In plane $\mathcal{P}$, lines m and n intersect at point A. If line k is perpendicular to line m and line n at point A, then line k is
(1) contained in plane $\mathcal{P}$ (3) perpendicular to plane $\mathcal{P}$
(2) parallel to plane $\mathcal{P}$ (4) skew to plane $\mathcal{P}$ 17 _____

GEOMETRY
June 2010

18. The accompanying diagram shows $\overline{AB}$ and $\overline{DE}$.

Which transformation will move $\overline{AB}$ onto $\overline{DE}$ such that point D is the image of point A and point E is the image of point B?
(1) $T_{3,-3}$ (3) $R_{90°}$
(2) $D_{\frac{1}{2}}$ (4) $r_{y=x}$

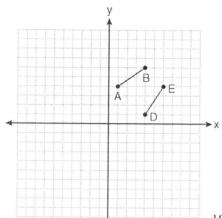

18 _____

19. In the accompanying diagram of circle O, chords $\overline{AE}$ and $\overline{DC}$ intersect at point B, such that $m\widehat{AC} = 36$ and $m\widehat{DE} = 20$.

What is $m\angle ABC$?
(1) 56   (3) 28
(2) 36   (4) 8

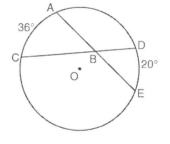

19 _____

20. The accompanying diagram shows the construction of a line through point P perpendicular to line m.

Which statement is demonstrated by this construction? 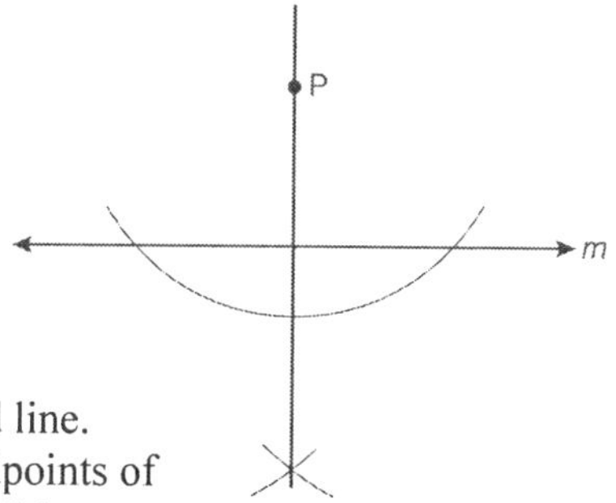
(1) If a line is parallel to a line that is perpendicular to a third line, then the line is also perpendicular to the third line.
(2) The set of points equidistant from the endpoints of a line segment is the perpendicular bisector of the segment.
(3) Two lines are perpendicular if they are equidistant from a given point.
(4) Two lines are perpendicular if they intersect to form a vertical line.

20 _____

21. What is the length, to the *nearest tenth*, of the line segment joining the points (–4, 2) and (146, 52)?
(1) 141.4   (2) 150.5   (3) 151.9   (4) 158.1   21 _____

22. What is the slope of a line perpendicular to the line whose equation is $y = 3x + 4$?
(1) $\frac{1}{3}$ 　　(2) $-\frac{1}{3}$ 　　(3) 3 　　(4) -3 　　22 _____

23. In the accompanying diagram of circle O, secant $\overline{AB}$ intersects circle O at D, secant $\overline{AOC}$ intersects circle O at E, $AE = 4$, $AB = 12$, and $DB = 6$.

What is the length of $\overline{OC}$?
(1) 4.5 　　(3) 9
(2) 7 　　(4) 14

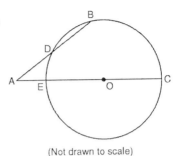
(Not drawn to scale)　　23 _____

24. The accompanying diagram shows a pennant in the shape of an isosceles triangle. The equal sides each measure 13, the altitude is $x + 7$, and the base is $2x$.

What is the length of the base?
(1) 5 　　(3) 12
(2) 10 　　(4) 24

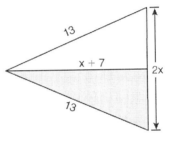
24 _____

25. In the accompanying diagram of $\triangle ABC$, $\overline{CD}$ is the bisector of $\angle BCA$, $\overline{AE}$ is the bisector of $\angle CAB$, and $\overline{BG}$ is drawn.

Which statement must be true?
(1) $DG = EG$
(2) $AG = BG$
(3) $\angle AEB \cong \angle AEC$
(4) $\angle DBG \cong \angle EBG$

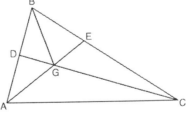
25 _____

26. In the accompanying diagram of circle O, chords $\overline{AD}$ and $\overline{BC}$ intersect at E. Which relationship must be true?
(1) $\triangle CAE \cong \triangle DBE$
(2) $\triangle AEC \sim \triangle BED$
(3) $\angle ACB \cong \angle CBD$
(4) $\overparen{CA} \cong \overparen{DB}$

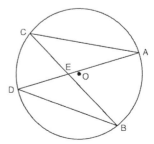
26 _____

GEOMETRY
June 2010

27. Two lines are represented by the equations $-\frac{1}{2}y = 6x + 10$ and $y = mx$. For which value of m will the lines be parallel?
(1) −12   (2) −3   (3) 3   (4) 12   27 _____

28. The coordinates of the vertices of parallelogram $ABCD$ are $A(-3, 2)$, $B(-2, -1)$, $C(4, 1)$, and $D(3, 4)$. The slopes of which line segments could be calculated to show that $ABCD$ is a rectangle?
(1) $\overline{AB}$ and $\overline{DC}$ (3) $\overline{AD}$ and $\overline{BC}$
(2) $\overline{AB}$ and $\overline{BC}$ (4) $\overline{AC}$ and $\overline{BD}$ 28 _____

Part II

Answer all 6 questions in this part. Each correct answer will receive 2 credits. Clearly indicate the necessary steps, including appropriate formula substitutions, diagrams, graphs, charts, etc. For all questions in this part, a correct numerical answer with no work shown will receive only 1 credit. All answers should be written in pen, except for graphs and drawings, which should be done in pencil. [12]

29. Tim is going to paint a wooden sphere that has a diameter of 12 inches. Find the surface area of the sphere, to the *nearest square inch*.

30. In the accompanying diagram of $\triangle ABC$, $\overline{DE}$ is a midsegment of $\triangle ABC$, $DE = 7$, $AB = 10$, and $BC = 13$. Find the perimeter of $\triangle ABC$.

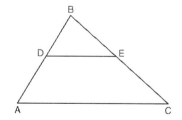

31. In right △DEF, m∠D = 90 and m∠F is 12 degrees less than twice m∠E. Find m∠E.

32. Triangle XYZ, shown in the diagram below, is reflected over the line $x = 2$. State the coordinates of △X'Y'Z', the image of △XYZ.

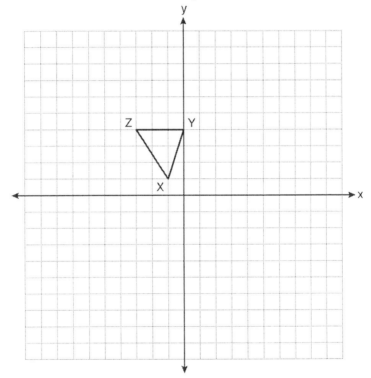

GEOMETRY
June 2010

33. Two lines, $\overleftrightarrow{AB}$ and $\overleftrightarrow{CRD}$, are parallel and 10 inches apart. Sketch the locus of all points that are equidistant from $\overleftrightarrow{AB}$ and $\overleftrightarrow{CRD}$ and 7 inches from point R. Label with an **X** each point that satisfies both conditions.

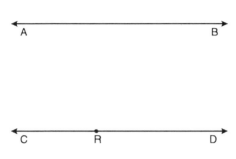

34. The base of a pyramid is a rectangle with a width of 6 cm and a length of 8 cm. Find, in centimeters, the height of the pyramid if the volume is 288 cm³.

GEOMETRY
June 2010
Part III

Answer all 3 questions in this part. Each correct answer will receive 4 credits. Clearly indicate the necessary steps, including appropriate formula substitutions, diagrams, graphs, charts, etc. For all questions in this part, a correct numerical answer with no work shown will receive only 1 credit. All answers should be written in pen, except for graphs and drawings, which should be done in pencil. [12]

35. Given: Quadrilateral $ABCD$ with $\overline{AB} \cong \overline{CD}$, $\overline{AD} \cong \overline{BC}$, and diagonal $\overline{BD}$ is drawn

    Prove: $\angle BDC \cong \angle ABD$

36. Find an equation of the line passing through the point (6, 5) and perpendicular to the line whose equation is $2y + 3x = 6$.

37. Write an equation of the circle whose diameter $\overline{AB}$ has endpoints $A(-4, 2)$ and $B(4, -4)$. [The use of the grid below is optional.]

GEOMETRY
June 2010
Part IV

Answer the question in this part. A correct answer will receive 6 credits. Clearly indicate the necessary steps, including appropriate formula substitutions, diagrams, graphs, charts, etc. A correct numerical answer with no work shown will receive only 1 credit. The answer should be written in pen. [6]

38. In the diagram below, quadrilateral $STAR$ is a rhombus with diagonals $\overline{SA}$ and $\overline{TR}$ intersecting at E. $ST = 3x + 30$, $SR = 8x - 5$, $SE = 3z$, $TE = 5z + 5$, $AE = 4z - 8$, m$\angle RTA = 5y - 2$, and m$\angle TAS = 9y + 8$. Find SR, RT, and m$\angle TAS$.

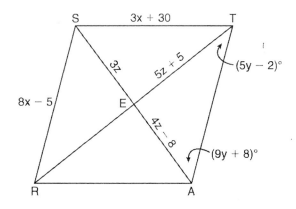

Collinear — on the same line

GEOMETRY
August 2010
Part I

Answer all 28 questions in this part. Each correct answer will receive 2 credits. No partial credit will be allowed. For each question, write on the space provided the numeral preceding the word or expression that best completes the statement or answers the question. [56]

1. In the accompanying diagram, $\triangle ABC \cong \triangle XYZ$. Which two statements identify corresponding congruent parts for these triangles? 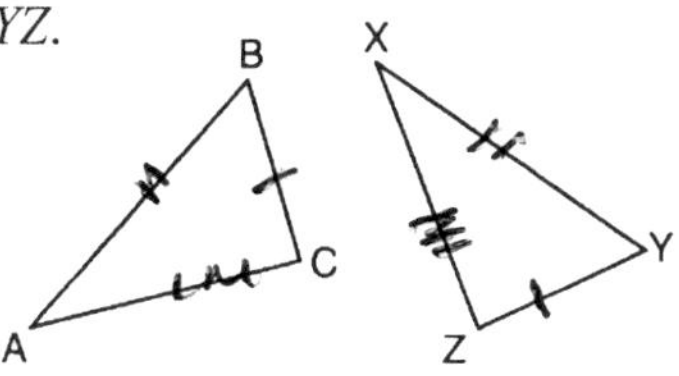
   (1) $\overline{AB} \cong \overline{XY}$ and $\angle C \cong \angle Y$
   (2) $\overline{AB} \cong \overline{YZ}$ and $\angle C \cong \angle X$
   (3) $\overline{BC} \cong \overline{XY}$ and $\angle A \cong \angle Y$
   (4) $\overline{BC} \cong \overline{YZ}$ and $\angle A \cong \angle X$

2. A support beam between the floor and ceiling of a house forms a 90° angle with the floor. The builder wants to make sure that the floor and ceiling are parallel. Which angle should the support beam form with the ceiling?
   (1) 45°   (2) 60°   (3) 90°   (4) 180°

3. In the accompanying diagram, the vertices of $\triangle DEF$ are the midpoints of the sides of equilateral triangle ABC, and the perimeter of $\triangle ABC$ is 36 cm. What is the length, in centimeters, of $\overline{EF}$?
   (1) 6   (2) 12   (3) 18   (4) 4

4. What is the solution of the following system of equations?
   $$y = (x + 3)^2 - 4$$
   $$y = 2x + 5$$
   (1) $(0, -4)$
   (2) $(-4, 0)$
   (3) $(-4, -3)$ and $(0, 5)$
   (4) $(-3, -4)$ and $(5, 0)$

5. One step in a construction uses the endpoints of $\overline{AB}$ to create arcs with the same radii. The arcs intersect above and below the segment. What is the relationship of $\overline{AB}$ and the line connecting the points of intersection of these arcs?
   (1) collinear   (2) congruent   (3) parallel   (4) perpendicular

6. If $\triangle ABC \sim \triangle ZXY$, $m\angle A = 50$, and $m\angle C = 30$, what is $m\angle X$?
   (1) 30   (2) 50   (3) 80   (4) 100

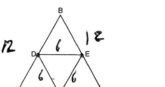

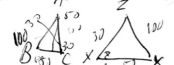

7. In the accompanying diagram of △AGE and △OLD, ∠GAE ≅ ∠LOD, and $\overline{AE} \cong \overline{OD}$. To prove that △AGE and △OLD are congruent by SAS, what other information is needed? 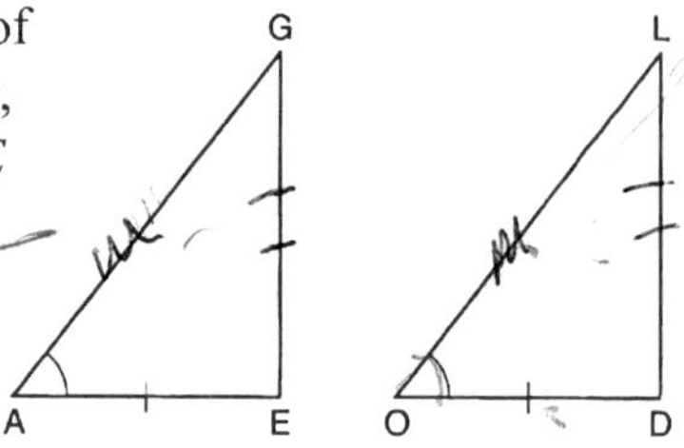
(1) $\overline{GE} \cong \overline{LD}$ (3) ∠AGE ≅ ∠OLD
(2) $\overline{AG} \cong \overline{OL}$ (4) ∠AEG ≅ ∠ODL

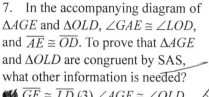
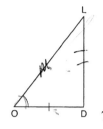
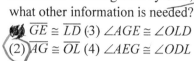

8. Point A is not contained in plane B. How many lines can be drawn through point A that will be perpendicular to plane B?
(1) one   (2) two   (3) zero   (4) infinite

9. The equation of a circle is $x^2 + (y - 7)^2 = 16$. What are the center and radius of the circle?
(1) center = (0, 7); radius = 4   (3) center = (0, –7); radius = 4
(2) center = (0, 7); radius = 16   (4) center = (0, –7); radius = 16

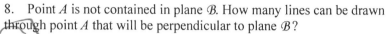

10. What is an equation of the line that passes through the point (7, 3) and is parallel to the line $4x + 2y = 10$?
(1) $y = \frac{1}{2}x - \frac{1}{2}$ (2) $y = -\frac{1}{2}x + \frac{13}{2}$ (3) $y = 2x - 11$ (4) $y = -2x + 17$

11. In △ABG, AB = 7, BC = 8, and AC = 9. Which list has the angles of △ABG in order from smallest to largest?
(1) ∠A, ∠B, ∠C
(2) ∠B, ∠A, ∠C
(3) ∠C, ∠B, ∠A
(4) ∠C, ∠A, ∠B

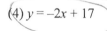

12. Tangents $\overline{PA}$ and $\overline{PB}$ are drawn to circle O from an external point, P, and radii $\overline{OA}$ and $\overline{OB}$ are drawn. If m∠APB = 40, what is the measure of ∠AOB?
(1) 140°   (2) 100°   (3) 70°   (4) 50°

13. What is the length of the line segment with endpoints A(–6, 4) and B(2, –5)?
(1) $\sqrt{13}$ (2) $\sqrt{17}$ (3) $\sqrt{72}$ (4) $\sqrt{145}$

14. The lines represented by the equations $y + \frac{1}{2}x = 4$ and $3x + 6y = 12$ are
(1) the same line
(2) parallel
(3) perpendicular
(4) neither parallel nor perpendicular

15. A transformation of a polygon that always preserves both length and orientation is
(1) dilation   (2) translation   (3) line reflection   (4) glide reflection

GEOMETRY
August 2010

16. In which polygon does the sum of the measures of the interior angles equal the sum of the measures of the exterior angles?
(1) triangle   (2) hexagon   (3) octagon   (4) quadrilateral   16_____

17. In the accompanying diagram of circle O, chords $\overline{AB}$ and $\overline{CD}$ intersect at E. If $CE = 10$, $ED = 6$, and $AE = 4$, what is the length of $\overline{EB}$?
(1) 15   (3) 6.7
(2) 12   (4) 2.4

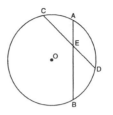

17_____

18. In the accompanying diagram of $\triangle ABC$, medians $\overline{AD}$, $\overline{BE}$, and $\overline{CF}$ intersect at G. If $CF = 24$, what is the length of $\overline{FG}$?
(1) 8   (2) 10   (3) 12   (4) 16

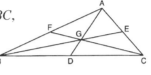

18_____

19. If a line segment has endpoints $A(3x + 5, 3y)$ and $B(x - 1, -y)$, what are the coordinates of the midpoint of $\overline{AB}$?
(1) $(x + 3, 2y)$ (2) $(2x + 2, y)$ (3) $(2x + 3, y)$ (4) $(4x + 4, 2y)$ 19_____

20. If the surface area of a sphere is represented by 144π, what is the volume in terms of π?
(1) 36π (2) 48π (3) 216π (4) 288π 20_____

21. Which transformation of the line $x = 3$ results in an image that is perpendicular to the given line?
(1) $r_{x\text{-axis}}$ (2) $r_{y\text{-axis}}$ (3) $r_{y=x}$ (4) $r_{x=1}$ 21_____

22. In the accompanying diagram of regular pentagon $ABCDE$, $\overline{EB}$ is drawn. What is the measure of $\angle AEB$?
(1) 36°   (3) 72°
(2) 54°   (4) 108°

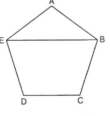

22_____

23. $\triangle ABC$ is similar to $\triangle DEF$. The ratio of the length of $\overline{AB}$ to the length of $\overline{DE}$ is 3:1. Which ratio is also equal to 3:1?

(1) $\dfrac{m\angle A}{m\angle D}$ (3) $\dfrac{\text{area of } \triangle ABC}{\text{area of } \triangle DEF}$

(2) $\dfrac{m\angle B}{m\angle F}$ (4) $\dfrac{\text{perimeter of } \triangle ABC}{\text{perimeter of } \triangle DEF}$

23_____

GEOMETRY
August 2010

24. What is the slope of a line perpendicular to the line whose equation is $2y = -6x + 8$?
(1) -3 (2) $\frac{1}{6}$ (3) $\frac{1}{3}$ (4) -6 24_____

25. In the accompanying diagram of circle C, $m\overset{\frown}{QT} = 140$ and $m\angle P = 40$. What is $m\overset{\frown}{RS}$?
(1) 50   (3) 90
(2) 60   (4) 100

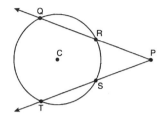

25_____

26. Which statement is logically equivalent to "If it is warm, then I go swimming"?
(1) If I go swimming, then it is warm.
(2) If it is warm, then I do not go swimming.
(3) If I do not go swimming, then it is not warm.
(4) If it is not warm, then I do not go swimming.   26_____

27. In the accompanying diagram of $\triangle ACT$, $\overleftrightarrow{BE} \parallel \overleftrightarrow{AT}$. If $CB = 3$, $CA = 10$, and $CE = 6$, what is the length of $\overline{ET}$?
(1) 5   (3) 20
(2) 14   (4) 26

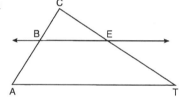

27_____

28. Which geometric principle is used in the construction shown below?
(1) The intersection of the angle bisectors of a triangle is the center of the inscribed circle.
(2) The intersection of the angle bisectors of a triangle is the center of the circumscribed circle.
(3) The intersection of the perpendicular bisectors of the sides of a triangle is the center of the inscribed circle.
(4) The intersection of the perpendicular bisectors of the sides of a triangle is the center of the circumscribed circle.

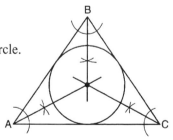

28_____

GEOMETRY
August 2010
Part II

Answer all 6 questions in this part. Each correct answer will receive 2 credits. Clearly indicate the necessary steps, including appropriate formula substitutions, diagrams, graphs, charts, etc. For all questions in this part, a correct numerical answer with no work shown will receive only 1 credit. [12]

29. The accompanying diagram shows isosceles trapezoid $ABCD$ with $\overline{AB} \parallel \overline{DC}$ and $\overline{AD} \cong \overline{BC}$.
If $m\angle BAD = 2x$ and $m\angle BCD = 3x + 5$, find $m\angle BAD$.

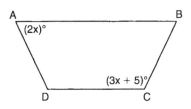

30. A right circular cone has a base with a radius of 15 cm, a vertical height of 20 cm, and a slant height of 25 cm. Find, in terms of π, the number of square centimeters in the lateral area of the cone.

31. In the accompanying diagram of $\triangle HQP$, side $\overline{HP}$ is extended through P to T, $m\angle QPT = 6x + 20$, $m\angle HQP = x + 40$, and $m\angle PHQ = 4x - 5$. Find $m\angle QPT$.

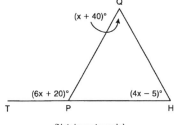

(Not drawn to scale)

GEOMETRY
August 2010

32. On the line segment below, use a compass and straightedge to construct equilateral triangle ABC. [Leave all construction marks.]

33. In the diagram below, car A is parked 7 miles from car B. Sketch the points that are 4 miles from car A and sketch the points that are 4 miles from car B. Label with an **X** all points that satisfy both conditions.

Car A

Car B

34. Write an equation for circle O shown on the accompanying graph.

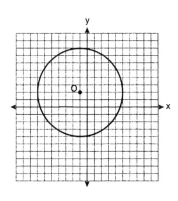

GEOMETRY
August 2010
Part III

Answer all 3 questions in this part. Each correct answer will receive 4 credits. Clearly indicate the necessary steps, including appropriate formula substitutions, diagrams, graphs, charts, etc. For all questions in this part, a correct numerical answer with no work shown will receive only 1 credit. [12]

35. In the accompanying diagram of quadrilateral $ABCD$ with diagonal $\overline{BD}$, m$\angle A$ = 93. m$\angle ADB$ = 43. m$\angle C$ = 3x + 5, m$\angle BDC$ = x + 19, and m$\angle DBC$ = 2x + 6. Determine if $\overline{AB}$ is parallel to $\overline{DC}$. Explain your reasoning.

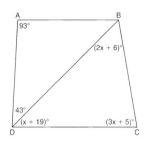

36. The coordinates of the vertices of $\triangle ABC$ are $A(1, 3)$, $B(-2, 2)$, and $C(0, -2)$, On the grid below, graph and label $\triangle A''B''C''$, the result of the composite transformation $D_2 \circ T_{3,-2}$. State the coordinates of A'', B'', and C''.

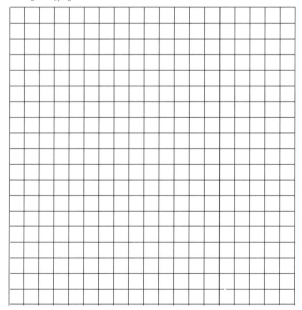

GEOMETRY
August 2010

37. In the accompanying diagram, △*RST* is a 3-4-5 right triangle. The altitude, *h*, to the hypotenuse has been drawn. Determine the length of *h*.

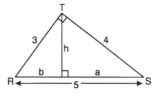

Part IV

Answer the question in this part. A correct answer will receive 6 credits. Clearly indicate the necessary steps, including appropriate formula substitutions, diagrams, graphs, charts, etc. A correct numerical answer with no work shown will receive only 1 credit. [6]

38. Given: Quadrilateral *ABCD* has vertices *A*(–5, 6), *B*(6, 6), *C*(8, –3), and *D*(–3, –3).

    Prove: Quadrilateral *ABCD* is a parallelogram but is neither a rhombus nor a rectangle. [The use of the grid below is optional.]

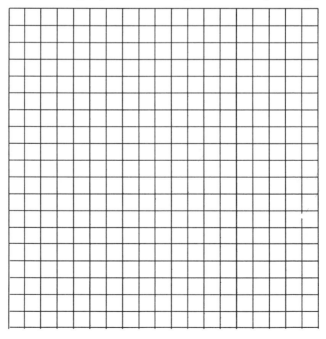